The Physics of Superionic Conductors and Electrode Materials

NATO Advanced Science Institutes Series

A series of edited volumes comprising multifaceted studies of contemporary scientific issues by some of the best scientific minds in the world, assembled in cooperation with NATO Scientific Affairs Division.

This series is published by an international board of publishers in conjunction with NATO Scientific Affairs Division

A Life Sciences **B Physics**	Plenum Publishing Corporation New York and London
C Mathematical and Physical Sciences	D. Reidel Publishing Company Dordrecht, Boston, and London
D Behavioral and Social Sciences **E Applied Sciences**	Martinus Nijhoff Publishers The Hague, Boston, and London
F Computer and Systems Sciences **G Ecological Sciences**	Springer Verlag Heidelberg, Berlin, and New York

Recent Volumes in Series B: Physics

Volume 87 —Relativistic Effects in Atoms, Molecules, and Solids
edited by G. L. Malli

Volume 88 —Collective Excitations in Solids
edited by Baldassare Di Bartolo

Volume 89a—Electrical Breakdown and Discharges in Gases: Fundamental Processes and Breakdown
edited by Erich E. Kunhardt and Lawrence H. Luesen

Volume 89b—Electrical Breakdown and Discharges in Gases: Macroscopic Processes and Discharges
edited by Erich E. Kunhardt and Lawrence H. Luessen

Volume 90 —Molecular Ions: Geometric and Electronic Structures
edited by Joseph Berkowitz and Karl-Ontjes Groeneveld

Volume 91 —Integrated Optics: Physics and Applications
edited by S. Martellucci and A. N. Chester

Volume 92 —The Physics of Superionic Conductors and Electrode Materials
edited by John W. Perram

The Physics of Superionic Conductors and Electrode Materials

Edited by

John W. Perram

Odense University
Odense, Denmark

Plenum Press
New York and London
Published in cooperation with NATO Scientific Affairs Division

Proceedings of a NATO Advanced Study Institute on
The Physics of Superionic Conductors and Electrode Materials,
held August 4–22, 1980,
at Odense University, Odense, Denmark

Library of Congress Cataloging in Publication Data

NATO Advanced Study Institute on the Physics of Superionic Conductors and Electrode Materials (1980: Odense universiteit)
The physics of superionic conductors and electrode materials.

(NATO advanced science institutes series. Series B, Physics; v. 92)
"Published in cooperation with NATO Scientific Affairs Division."
Bibliography: p.
Includes index.
1. Superionic conductors—Congresses. 2. Electrodes—Congresses. I. Perram, John W. II. North Atlantic Treaty Organization. Scientific Affairs Division. III. Title. IV. Series.
QC717.N37 1980 537.6′2 83-2332
ISBN-13: 978-1-4684-4492-6 e-ISBN-13: 978-1-4684-4490-2
DOI: 10.1007/978-1-4684-4490-2

Softcover reprint of the hardcover 1st edition 1983
A Division of Plenum Publishing Corporation
233 Spring Street, New York, N.Y. 10013

PREFACE

The following chapters present most of the lectures delivered at the NATO Advanced Studies Institute on "The Physics of Superionic Conductors and Electrode Materials", held at Odense University's Mathematics Department between the 4th and 22nd of August, 1980. The aim of the organizing committee was to present in a rather detailed fashion the most recent advances in the computational mathematics and physics of condensed matter physics and to see how these advances could be applied to the study of ionically conducting solids. The first half of the meeting was mainly taken up with lectures. In the second week, working groups on the various aspects were set up, the students joining these groups being helped in the implementation of the lecture material. The leaders of these groups deserve special mention for the tremendous effort they put into this aspect of the meeting, particularly:

Dr. Aneesur Rahman (Molecular Dynamics group)
Dr. Fred Horne (Ion Transport group)
Drs. Nick Quirke and David Adams (Monte Carlo methods)
Dr. Heinz Schulz (Diffraction group)
Dr. John Harding (Defect Calculations group)

The Molecular Dynamics group achieved a certain amount of notoriety within the University by appearing to live in the terminal room. The computing aspect of the meeting would have been much less successful without the assistance of Jan Kennett, of the Odense Datacenter, who gave a talk to participants on the computer's operating system and was always available to assist with programming and system problems. The Datacenter also assisted the meeting with a generous allocation of computer time.

The meeting also owes a debt of gratitude to the University's Rektor, Dr. Aage Trommer, for ensuring that the University's facilities were available to us, and to the Mayor of Odense, Mr. Werner Dalskov and the City for their hospitality during the meeting.

Finally, the Director would like to express his gratitude to the participants, whose appetite for work and tolerance of red cabbage ensured the success of the meeting.

CONTENTS

INTRODUCTION

John W. Perram

Matematisk Institut
Odense Universitet
DK 5230, Odense, Denmark

The price of oil and the political instabilities associated with its sources of supply have led technologists to look at other forms of portable energy, the most obvious candidate being electrochemical energy stored in batteries. This has generated, in turn, interest in the scientific investigation of electrolytes other than aqueous ionic solutions as candidates for inclusion in the conducting phase of advanced batteries.

It has been known for a long time that certain crystals become ionic conductors above a transition temperature. Most of the well known cases, such as silver iodide, the alkaline earth fluorites, and lithium nitride have little potential technical application themselves. However, in studying them, we are seeking to understand how they work, with a view to applying our knowledge of mechanism to devising new materials with technological applications. There are far too many crystals for a project of exhaustive conductivity testing to have much chance of success.

It is to the elucidation of mechanism that this workshop and this volume are directed. The chapters can be regarded as a source book in methodology. There are any number of excellent reviews and books on the state of the art in the field of ionic conductors. However, the scientist who wishes to understand how these materials work will find it difficult to get access to the necessary theoretical background. It is thus hoped that the budding theoretician, in this and other areas, will find these lectures useful. Some of the topics require a fair amount of mathematical expertise for their understanding, although they contain enough detail for the reader to be able to follow the argument. For this, the editor makes no apology: the theoretical problems in this area are not easy.

In this task, no amount of experimental work can tell us anything in the absence of a proper theoretical framework. However, theories must be grounded in reality, and the early lectures of the meeting were directed towards phenomenology. Of these, the first, by Professor Schulz, gives a clear and elegant elucidation of what can be learnt by diffraction studies, and how the rationale for conducting experiment is to discover mechanism.

In Chapter 2, Dr. Harding discusses the conduction mechanism from a quasi-thermodynamic point of view, describing the so-called "hopping model". In later lectures, Dr. Catlow developed these ideas and described the methodology of defect calculations in ionic crystals, discussing in a general way the techniques and assumptions used in developing computer codes of the HADES and PLUTO families.

In the next two chapters, Dr. Smith gives a clear, detailed exposition of the calculation of lattice sums in both perfect and defective crystals. These techniques turn out to be crucial for the carrying out and interpretation of simulations of all types.

In Chapter 5, Professor Cotterill gives an overview of the application of computer simulation techniques to the study of the transition from ordered solid to disordered fluid. This lecture is included because it sheds light on how solids melt, and whether sub-lattice melting could be responsible for ionic conductivity.

In Chapter 6, Drs. Rahman and Vashishta describe in admirable detail how one goes about performing a molecular dynamics simulation of ionic systems in general and superionic conductors in particular. This article reviews all the main molecular dynamics work and shows how the method may be extended to operate at constant pressure and how structural phase transitions may be studied. It will undoubtedly become a standard reference work in this area.

In Chapter 7, Dr. Smith addresses the difficult problem of including the effects of electronic polarizability in the calculations. It should be stressed that all molecular dynamics calculations treat the ions as non-polarizable. Here one obtains a closer approximation to reality at the cost of acquiring a truly many-body interaction.

In Chapter 8, Dr. de Leeuw describes the role of boundary conditions in ionic simulations. He shows how these act in a very subtle way to determine the calculated conductivity. In so doing, he answers one of the long-standing puzzles as to why a computer simulation of a molten salt gives a D.C. conductivity at all.

In Chapter 9, Dr. Adams introduces the second method of computer simulation, the so-called Monte Carlo method. He discusses how one obtains structural and thermodynamic information by sampling the phase space rather than evaluating time averages. Again the exposition is clear and detailed and all significant variants of the method are described.

In Chapter 10, Dr. Quirke shows how the Monte Carlo method may be used to calculate free energies, the one quantity inaccessible by the molecular dynamics technique. He describes recent work on the umbrella sampling method and discusses how these methods may be applied to systems of interest here.

In Chapter 11, Drs. Hiwatari and Ueda present details of a Monte Carlo study of calcium fluoride and silver iodide. Their work supplements the molecular dynamics studies of the same systems in Chapter 6.

In Chapter 12, Drs. Berendsen and van Gunsteren take up a problem which has recently attracted a lot of interest in the field, namely the question of polymeric electrolytes. It seems that substances such as polyethylene oxide which have a relatively high dielectric constant can act as solvents for small inorganic ions. Such plastic substances have more desirable materials properties than either solids or liquids. The authors describe how the molecular dynamics method can be applied to such systems, using the method of constraint dynamics pioneered by them.

In Chapter 13, the same authors describe the method of stochastic dynamics, a type of hybrid technique. They show how the forces from the most rapid motions in the system may be treated in a random fashion while focussing on the slower dynamic processes and thus saving a lot of computer time.

Complementary to understanding the microscopic dynamics of of conductors is the problem of predicting fluxes and fields in macroscopic samples of them under various conditions. In Chapter 14, Dr. Horne gives a survey of the application of irreversible thermodynamics to mass transport in ionic conductors. He particularly stresses the important role of the Onsager coefficients and reviews recent exciting work on the application of finite element numerical techniques to the solution of electrodiffusion equations. A particular topic of discussion in the ion transport group was the question of proper boundary conditions for the migration of ions across phase boundaries. Chapter 15 describes briefly some progress made on this problem during and after the symposium.

This work would not be complete without the editor expressing his enormous debt of gratitude to Lisbeth Larsen, who retyped several of the manuscripts.

RELATIONS BETWEEN CRYSTAL STRUCTURES AND IONIC CONDUCTIVITY

Heinz Schulz

Max-Planck-Institut fur Festkorperforschung
Heisenbergstr. 1, D-7000 Stuttgart 80
Federal Republic of Germany

ABSTRACT

High ionic conductivity combined with very low electronic conductivity has been observed in several crystals. Crystal structure analysis and the interpretation of diffuse X-ray scattering have been successfully applied in the investigation of the ionic conduction mechanism in such crystals. Ions may move easily and rapidly in crystals if their ionic radius fits into the cross sections of the conducting paths and if these paths connect regular lattice sites. The crystal field potential shows a rather weak variation along these diffusion paths. The regular lattice sites form local but shallow potential minima in such paths. These sites are not allowed to be completely occupied. The amount of underoccupancy ranges from about 1% to more than 50%.

Introduction

Ions can move through a solid under the influence of an electrical field. Thereby mass and charge are transported through the solid in contrast to electronic conductivity which transports almost only electrical charge. Ionic conductivity can be observed by so-called transference measurements, in which material from one side of a solid can be transported to the other side of the solid by means of an electrical field. A transference number of one is obtained and Faradays law is obeyed if the charge calculated from the transported mass is equal to the charge calculated from the applied electrical current and the time the experiment lasted.

Ionic conductivity in solids was already reported by Faraday[1] more than one hundred years ago. In 1884 Warburg[2] described the migration of sodium through a glass and its precipation on the surface of the glass when a direct current flowed through the glass. Nernst reported in 1899 high ionic conductivity in mixed oxides at high temperatures[3].

In 1914, Tubandt and Lorenz[4] detected the extraordinarily high silver conductivity of the alpha-phase of silver iodide which occurs above 147°C. Its conductivity ranges between 1.2 and 2.6 $(\Omega cm)^{-1}$which is comparable to the best conducting liquid electrolytes. These results could not be explained for decades. The understanding of solid electrolytes matured with the development of the crystal structure analysis and especially with the investigation of disordered materials. The first ideas on the ionic conductivity of crystalline materials were derived from the crystal structure investigation of α-AgI which was carried out in 1938 with polycrystalline samples by Strock[5].

It is known that materials with high ionic conductivity are characterized by common structural features. An ionic conductor belongs to this group if its conductivity is larger than 5.10^{-2} $(\Omega cm)^{-1}$ and if its electronic conductivity is at least two orders of magnitude lower. They are designed in the literature as fast, optimized or superionic conductors. A selection of such conductors is shown in Fig. 1.

Structural Aspects

The structure determination of α-AgI by Strock[5] showed that only the I^- ions form an ordered bcc lattice (Fig. 2). The two Ag ions per elementary cell seemed to be distributed over 42 sites with different I coordinations in the form of octahedra (6x), tetrahedra (12x) and trigonal pyramids (24x). The distinction between a regular Ag-site and an interstitial Ag-site was not possible and therefore it was concluded the Ag sublattice is in a molten state. Since then the argument has often been used that high conductivity can only exist in crystals with partly molten lattices. However, in a liquid anions and cations are mobile and can rearrange themselves in order to reduce potential differences. In a crystal with a so-called molten sublattice, the mobile ions have to move through the periodical potential generated by the ions forming the ordered lattice.

In this connection, it is interesting to have a look at the crystal structure of ß-AgI (Fig. 3). In ß-AgI which has low ionic conduction the I sublattice forms tetrahedral and octahedral voids. The Ag ions occupy only tetrahedral sites in an ordered manner.The unoccupied octahedra are connected with each other and with the tetrahedra by common faces. The octahedra form rows parallel to the c-axis. This seems to be the ideal arrangement for an ionic conductor, if only geometrical features are considered. Thus an

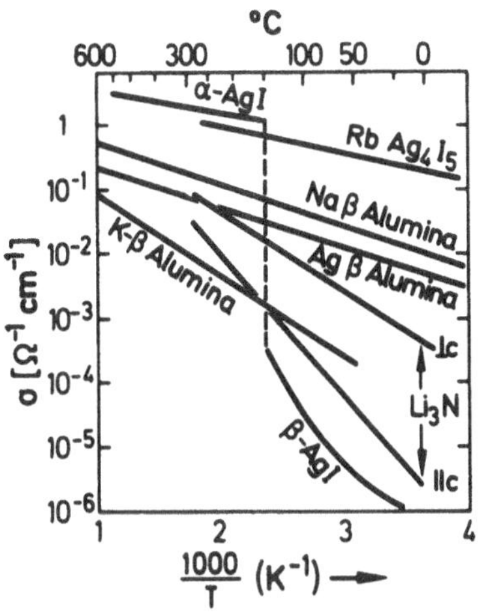

Fig. 1. Ionic conductivity of Silver Iodide, ß-Alumina with different conducting ions, Rubidium Silver Iodide and Lithium Nitride as a function of temperature. For Li_3N the conductivity is given parallel and perpendicular to the hexagonal c-axis. $RbAg_4I_5$ shows the highest ionic conductivity at room temperature reported up to now.

Ag ion may jump from a tetrahedral site into an octahedron and may move now through the crystal within a row of octahedra (Fig. 3). However, at 140°C only about 1% of the Ag ions occupy the octahedra. This result suggests that ionic conduction involving interstitial sites (Frenkel disorder) does not allow high ionic conduction. Only ion movement between regular lattice sites generates high conductivity values. In a more general way, jumps of ions between sites with similar site energies are required. Such jumps require an occupational disorder as found in the α-form of AgI. However, there exist many compounds with large degrees of occupational disorder, e.g. zeolites, which show only very low ionic conductivity. This discrepancy has been studied by conductivity measurements on ß-alumina. ß-alumina is a two-dimensional ionic conductor with the ideal composition $Na_2O.11\ Al_2O_3$. However, these materials contain usually more sodium and it is rather difficult to generate the stochiometric composition. The Na^+ ions can easily be exchanged with many other ions, e.g. Li^+, K^+, Rb^+, Ag^+ or Tl^+. The structure may be described by spinel(Al_2O_3) - blocks which are separated from each other by bridging oxygens (Fig. 4a). In this way layers perpendicular to hexagonal c-axis are generated which contain the mobile ions (Fig. 4b). The ions can move along a path in a honey-comb, the path being formed by the Beevers-Ross,

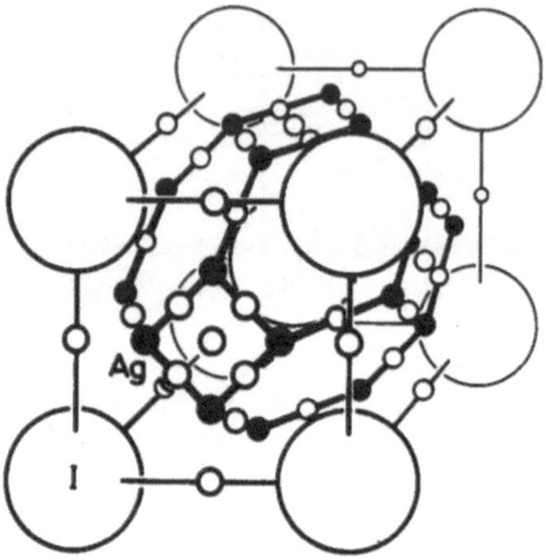

Fig. 2. Crystal structure of α-AgI.
The I ions form a bcc lattice. The silver distribution proposed by Strock (5) is shown by the small circles. A recent neutron diffraction experiment[31] revealed, that only the tetrahedral voids (black small circles) are occupied by the silver ions. These ions move along the heavy black lines through the lattice.

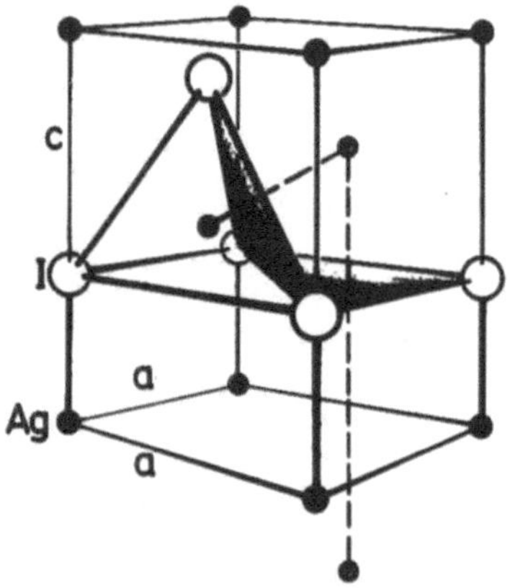

Fig. 3. Crystal structure of ß-AgI.
ß-AgI crystallizes in the wurtzite type structure. The figure shows a possible Ag-migration path: A silver ion jumps from a tetrahedral void into an unoccupied octahedra (first hatched position) and moves then parallel to the c-axis through the chain of unoccupied octahedra (second hatched position).

Anti-Beevers-Ross and mid-oxygen positions. The occupation probability of each site depends on the ion species and is also function of the temperature. Reviews on β-Alumina investigations have recently been published in references 6 and 7.

The conductivity of β-alumina depends on the sort of conducting ions as shown in Fig. 1. Huggins[8] has measured the activation enthalpy for different ions in β-Alumina (Fig. 5) where a pronounced minimum is seen for Na^+ β-alumina.

Repeating of these measurements under pressure showed a shift of the curve towards lower ionic radii (Fig. 5), i.e. the enthalpy of Li^+ motion decreased and the enthalpy of Tl^+ motion increased under pressure[9]. It was concluded from these measurements that the ionic radius of the mobile ion must fit into the geometrical dimensions of the conducting path. The motion through the crystal lattice is restricted for ions with large ionic radius. Ions with very small radius fall into holes of the crystal field potential. Ions with the optimal radius follow a smooth path with rather low variation of the potential along it. Similar results were obtained for α-AgI[10].

The following conclusions can be drawn from the above mentioned investigations of fast ionic conductors. Ions may move easily and rapidly in crystals with a high degree of positional disorder, if their ionic radius fits into the cross sections of the conducting paths and if these paths connect regular lattice sites with each other.

Thermal Vibrations and Diffusion Paths

The accuracy of structure investigations has been increased enormously during the last decade due to the development of automatic diffractometers and fast computers. Nowadays it is no problem to collect 10 000 reflections and to process this large amount of measurements using highly sophisticated Fourier methods, which need a lot of computer time. This information allows not only the determination of the geometrical features of the crystal structure, but also the dynamical properties of the crystals by means of Bragg intensities. However, one has to realize, that dynamical properties can only be derived in an indirect way from crystal structure investigations. X-ray diffraction generates only an average over time and space. The interaction between X-rays and atoms lasts 10^{-18} sec and a thermal vibration or a jump of an atom lasts 10^{-12}-10^{-13} sec. Therefore the X-ray beam gives only a snap-shot of a frozen-in crystal structure and these snap-shots are superimposed in X-ray diffraction experiments. Therefore crystal structure investigations based on X-ray diffraction data deal only with an average structure. This holds in a similar way for elastic neutron diffraction. The structure investigations allow one to calculate density maps in

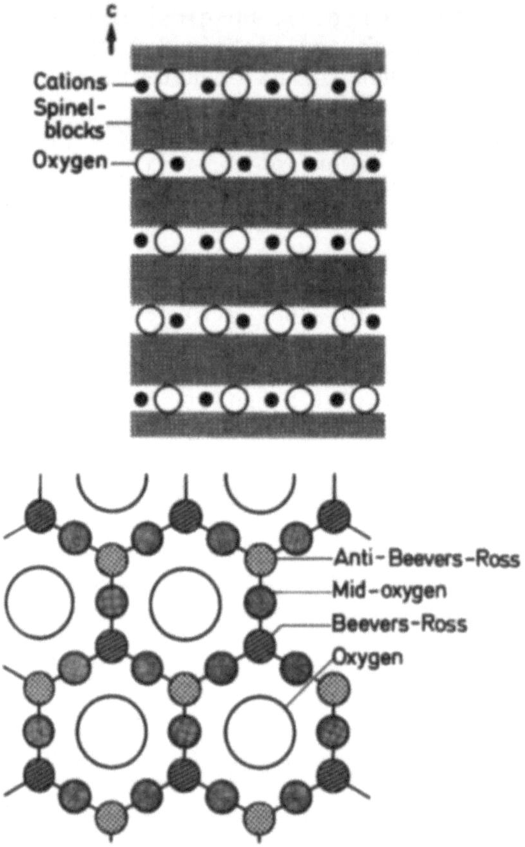

Fig. 4. Crystal structure of ß-Alumina.
a) Schematic view parallel to the c-axis.
The conducting layers are formed by bridging oxygens and the conducting ions. These conducting layers are separated from each other by spinel -(Al_2O_3)-blocks.
b) Section through the conducting layer.

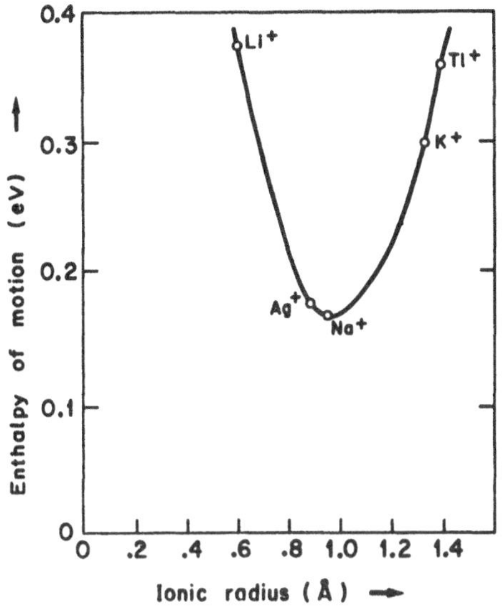

Fig. 5. Enthalpy of motion for ions with different ionic radius in ß-Alumina[8].

which the jumping or diffusing ions show up. The diffusion paths of the mobile ions can therefore be taken from the Fourier maps and in this way the ionic conduction mechanism can be derived from diffraction experiments.

This can be illustrated with the structure investigation of lithium nitride (Li_3N). Li_3N shows a high lithium conductivity [11,12] which is anisotropic (Fig. 1.) The crystal structure of Li_3N can be described as a sequence of Li and Li_2N layers perpendicular to the hexagonal c-axis (Fig. 6).

The nitrogens occupy the center of the elementary cell. They are surrounded by eight lithium ions in the shape of a hexagonal bi-pyramid. In a standard structure investigation a completely ordered lithium distribution was found[13]. More detailed investigagation of this structure[14,15] in the temperature range -120°C to to 630°C made it seem probable that the Li(2) ions (Fig. 6) show a weak unoccupancy of 1-2% up to 400°C. Above this temperature a rapid decrease of the occupation probability was observed. The Li(1) ions do not show a measurable underoccupancy. It follows, this material exhibits both, high ionic conductivity and a high degree of ordering.

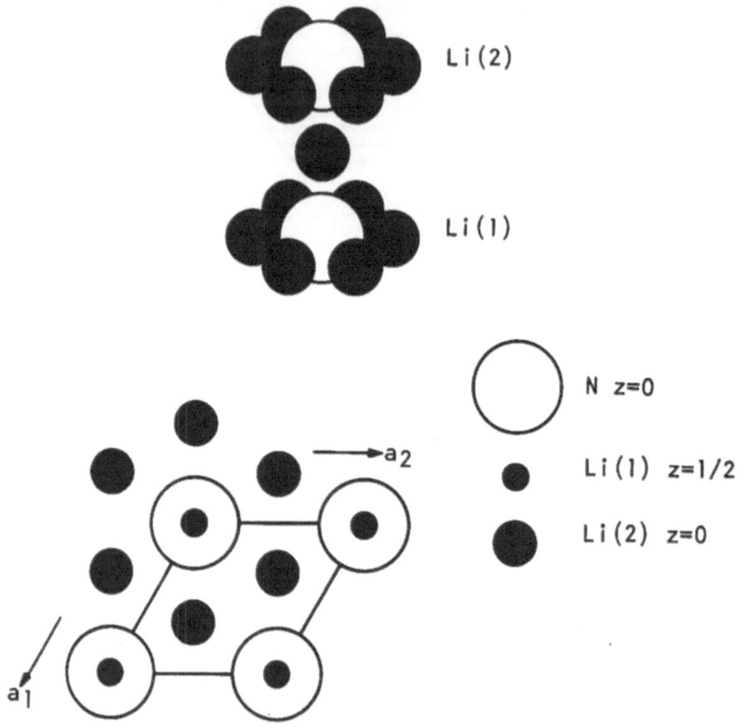

Fig. 6. Crystal structure of lithium nitride.

a) Projection along the hexagonal c-axis on the (a_1, a_2) plane.

b) Perspective drawing of the atomic arrangement.

The ionic conduction mechanism can be derived from Fourier difference maps. These maps show the differences between the observed electron density and a calculated electron density which is based on a refined structure model. For Li_3N the usual temperature factors and the above mentioned occupation probabilities were refined. The corresponding difference maps are displayed in Fig. 7 and 8.

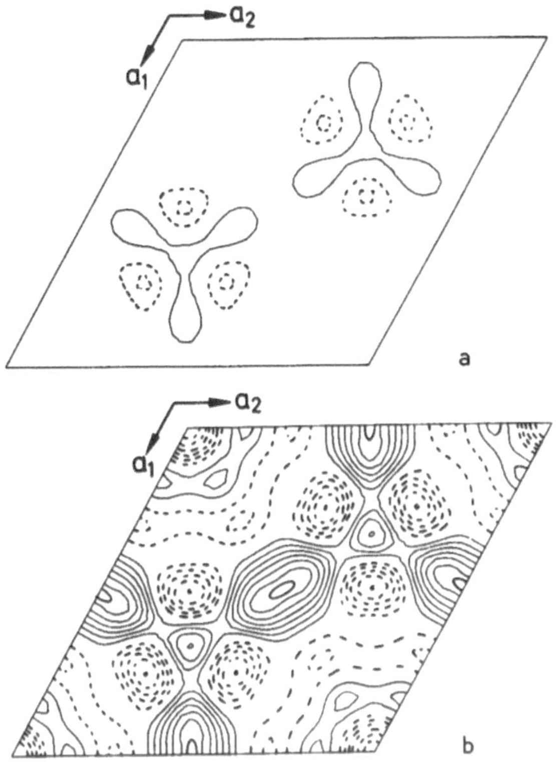

Fig. 7. Residual electron density of Li_3 in the Li(2) N plane at z=0. Positive and negative electron densities are shown by full and dotted lines. The lines are drawn at 0.03, 0.06,...e/A^3. a) 405°C b) 615°C

They show the developments of residual positive and negative densities around the Li(2) positions with increasing temperature. These positive densities grow together and form a two-dimensional net in the $Li(2)_2N$ plane along the Li(2)-Li(2) connection lines (Fig. 7). They suggest a diffusion path of mobile Li(2) ions within these planes. Similar conclusions can be drawn from Fig. 8. The droplike positive residual densities extend parallel to the c-axis with increasing temperature. The densities make it probable that the Li(2) ions can jump from a Li_2N layer to the neighbouring layers. These are either direct jumps (Li(2)-Li(2)) or indirect jumps (Li(2)-Li(1)-Li(2)). It follows that the Li(2) ions play the most important role in the ionic conduction of this material. The mobile ions move along paths as shown in Fig. 7 and 8. These paths form a three-dimensional network along which the Li(2) ions can move through the whole crystal. The structure investigations of this material have shown that high ionic conductivity is not necessarily connected with a high degree of occupational disorder. The underoccupation of the Li(2) position is so small that it is hard to detect it by a structure investigation.

Anharmonic Temperature Factors

The residual densities of Fig. 7 and 8 reveal that the used structure model is not able to describe the smeared densities which are typical for crystals with high ionic conduction. The structure model was based on the usual structure factor equation.

$$F(\underset{\sim}{h}) = \sum_{j=1}^{N} f_j(\underset{\sim}{h})\ T_j \exp(2\pi i\ h_m x_j^{m}) \tag{1}$$

N: Number of atoms

$\underset{\sim}{h}$ ≠,$(h_1 h_2 h_3)$: reciprocal lattice vector and its components

$\underset{\sim}{x}$ ≠,$(x^1 x^2 x^3)$: lattice vector and its components

f_j: atomic scattering factor of the jth atom

f_j: temperature factor of the jth atom

$h_m x^m = \sum_{m=1}^{3} h_m x^m$: (Einstein convention)

In routine structure investigations it is assumed that the atoms are located in harmonic potential wells (Fig. 9). Their thermal vibrations can then be described by mean square displacements $\overline{u^2}$. This can be done with the well known Debey-Waller Factor (in crystallography designated as isotropic temperature factor B). The displacements are described by an $\overline{u^2}$ sphere in this case. Frequently an anisotropic temperature factor is used which corresponds to an $\overline{u^2}$-ellipsoid The anisotropic temperature factor for an atom j may be written as:

$$T_j = \exp(-B_j^{mn} h_m h_n) \tag{2}$$

Isotropic temperature factor B and anisotropic temperature factor B_{mn} describe only thermal vibrations in a harmonic potential well which do not allow the atoms to leave this potential at any temperature (Fig. 9). This is not true for some ions in ionic conductors as, for example in lithium nitride. It is more probable that these ions occupy anharmonic potentials as shown in Fig. 9. The thermal vibrations of such ions have to be described with so called an-

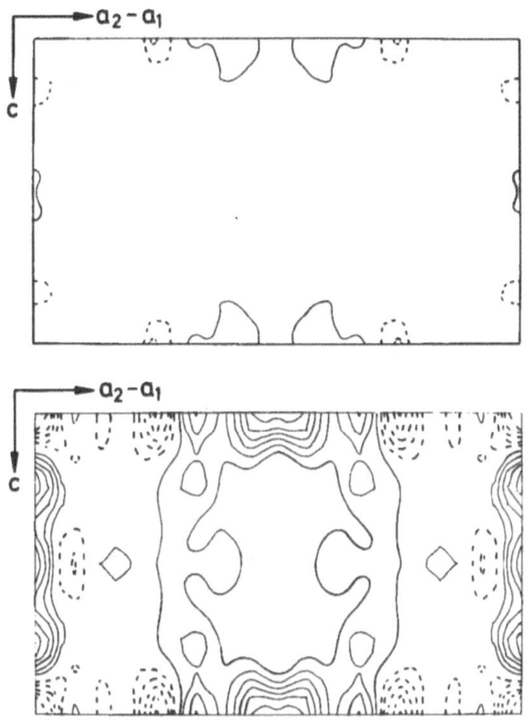

Fig. 8. Residual electron density of Li_3N in the (a_2-a_1)-c plane. The c-axis coincides with the N-Li-(1)-N connection line The (a_2-a_1) direction coincides with the Ni-Li(2)Li(2)-N-line. Positive and negative electron densities are shown by full and dotted lines. The lines are drawn at 0.03, 0.06...e/A^3.
a) 405°C b) 615°C

harmonic potentials. The relations between anharmonic potentials and the corresponding anharmonic temperature factors are illustrated in Fig. 10. It shows an anharmonic potential, which is seen by all the atoms occupying one atomic position. These atoms have different thermal excitations. For the whole crystal these excitations are controlled by the Boltzman statistics. Averaging over the whole crystal leads to distribution of atoms in this potential as indicated by the grey scale in Fig. 10. The superimposed distribution of all these atoms represents the probability density function of this atomic position. It is a function of the temperature and the form of the potential. The temperature factor of this atomic position is now given by the Fouriertransform F of the pdf:

$$T_j = F\ (pdf) \tag{3}$$

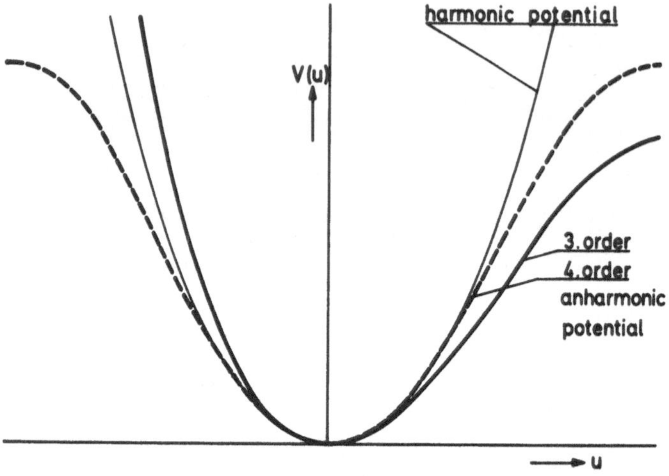

Fig. 9. Different potential forms.

Anharmonic temperature factors can be calculated, if a function exists, which describes an arbitrary pdf and which can be Fourier-transformed. There exists two series expansions, which fulfil these conditions of the Edgeworth expansion[16-18] and the Gram-Charlier expansion[17,18]. These series expansions can be used for all crystal symmetries and they have been applied for the determination of anharmonic temperature factors[19,20]. The Edgeworth expansion has the form:

$$pdf(\underset{\sim}{x}) = \exp\left(\frac{1}{3!} D_q D_r D_s c_j^{qrs} + \frac{1}{4!} D_q D_r D_s D_t d_j^{qrst} + \ldots\right) ph_j(x) \tag{4}$$

$$D_q = \frac{d}{dx_q}$$

$\exp(D_q)$: is defined by the Taylor expansion of the exponential function.

c,d,...: total symmetrical tensors of rank 3,4,...

$ph(\underset{\sim}{x})$: harmonic probability density function (Gaussian function).

$\frac{d}{dx_q} ph(\underset{\sim}{x}) \rightarrow$ Hermit polynoms

$ph(\underset{\sim}{x})$: Gaussian function

The Fouriertransform of (4) leads to the corresponding anharmonic temperature factor:

$$T_j(\underset{\sim}{h}) = \exp(\frac{1}{3!}(2\pi i)^3 c_j^{qrs} h_q h_r h_s + \frac{1}{4!}(2\pi i)^4 d_j^{qrst} h_q h_r h_s h_z + \ldots)\, Th_j(\underset{\sim}{h}) \qquad (5)$$

Th: harmonic temperature factor.

The corresponding equations for the Gram-Charlier expansion are:

$$pdf(\underset{\sim}{x}) = (1 + \frac{1}{3!} c_j^{pqr} D_p D_q D_r + \ldots)\, ph_j(\underset{\sim}{x}) \qquad (6)$$

$$T_j(\underset{\sim}{h}) = (1 + \frac{1}{3!}(2\pi i)^3 c_j^{pqr} h_p h_q h_r + \ldots)\, Th_j(\underset{\sim}{h}) \qquad (7)$$

Eq. (5) or (7) can be put into eq. (1). This structure factor equation can now be fitted to measured structure amplitudes by least squares methods. There exist relations between the coefficients of the tensors c, d, ..., which depend on the symmetry of the atomic position.

The influence of the anharmonic temperature factor on the accuracy of a structure refinement is now demonstrated with lithium nitride. Fig. 7a shows the residual density after the structure refinements with harmonic temperature factors. There exist areas with large positive and negative densities. The corresponding residual electron densities for refinements with anharmonic temperature factors are shown in Fig. 11a-c. The anharmonic temperature based on the Edgeworth expansion and applied up to 4th order does not reduce very much the residual densities (compare Figs. 7a and 11a). This is in considerable contrast to Gram-Charlier expansion (Fig. 11b). A nearly featureless difference density map is generated, if the Gram-Charlier formalism is applied up to the 6th order. (Fig. 11c), this means, a nearly perfect agreement between calculated and observed structure amplitudes has been achieved.

Probability density maps

The structure refinements with anharmonic temperature factors result in a number of coefficients of these temperature factors, which do not allow a straightforward interpretation. Furthermore, usually there exist strong corellations among these coefficients. The only way to overcome these difficulties is the calculation of the probability density maps by means of eq. (4) and (6). In this way all coefficients are combined, the corellations do not play a role now and the result is a function in real space, which allows a direct interpretation. The probability density map calculated

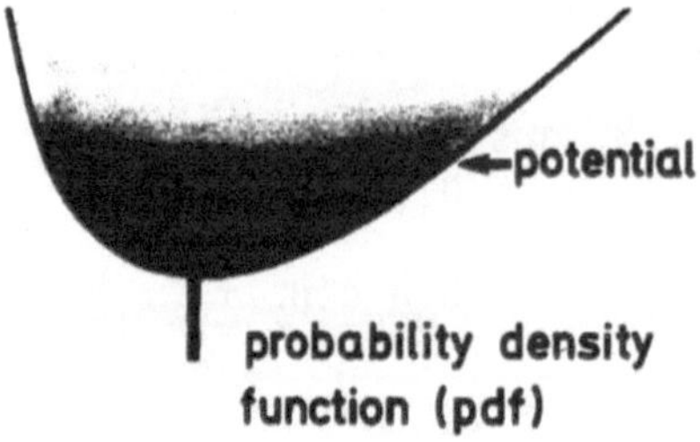

Fig. 10. Relations between an anharmonic potential, probabilities density function and anharmonic temperature factors. The shadowed part of the potential represents the ensemble of all ions which occupy this potential. The change from dark grey to light grey marks the decrease of the number of atoms excited to higher energies.

with eq. (6) with coefficients up to the 6th order shows Fig. 12. The coefficients were taken from the structure refinement, which resulted in the residual density map of Fig. 11c. The pronounced feature of Fig. 12 is the trigonal symmetry of pdf for the Li ions, which shows at one glance the strong deviation from a harmonic (Gaussian) pdf. The pdf of the nitrogen ion has a circle like symmetry. A hexagonal symmetry, although allowed, is not observable. However, also this pdf is not harmonic. It has a plate like shape. The pdf of Fig. 12 shows clearly the pronounced thermal motion of the Li(2) ions into the directions of their Li(2) neighbours, which is responsible for the high ionic conduction of Li_3N perpendicular to the hexagonal c-axis. Similar pdf maps can be calculated for any plane.

The high temperature phase of the fast ionic conductor Ag_3SI[30] shall now be used as a second demonstration for this type of crystal structure investigation of ionic conductors. This material belongs to the famous group of Ag ionic conductors AgI[5-12], Ag_2S[22], and Ag_3SI[23]. Their high temperature phases show very high silver ionic conduction. The structure consists of a I, S of (S,I) bcc framework. The Ag show a high degree of positional disorder.

The silver density of α-Ag_3SI (Fig. 15) looks like a completely delocalized silver distribution with very weak maxima. An excellent fit of a calculated density to the density of Fig. 15 was achieved by application of anharmonic temperature factors based on the Gram-Charlier expansion. Their coefficients have been used for the calculation of the Ag-pdf's of Fig. 16. The maximum density of

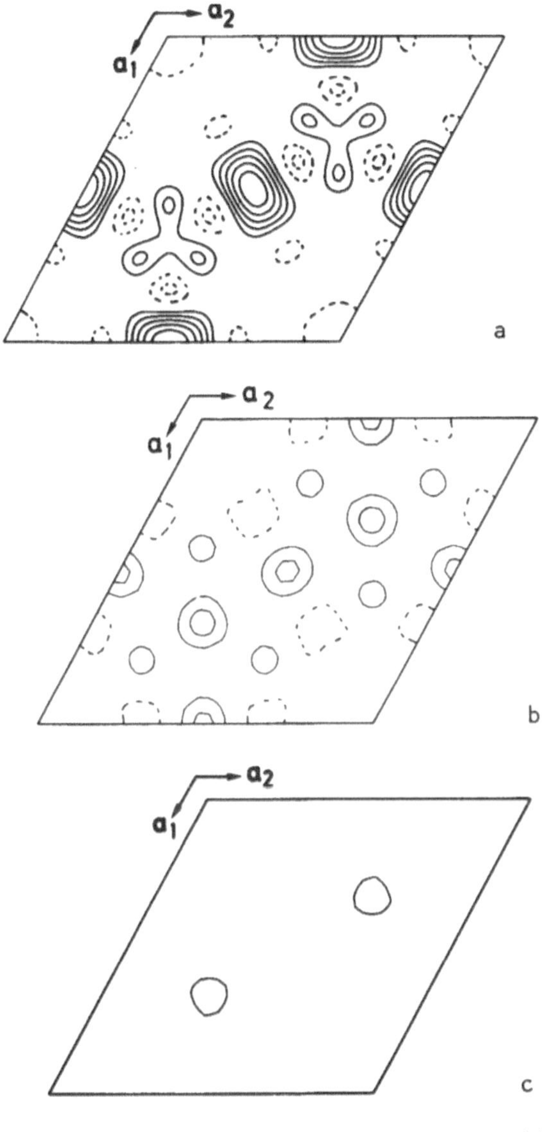

Fig. 11. Residual electron densities of Li_3N in $Li_2(2)$ N plane at z=0. Refinements with anharmonic temperature factors.

(a) Refinements based on Edgeworth expansion up to the 4th order (a);

(b) Refinement based on the Gram-Charlier expansion up to the 4th order;

(c) Refinement based on the Gram-Charlier expansion up to the 6th order.

Lines are drawn at the same levels as in Fig. 7

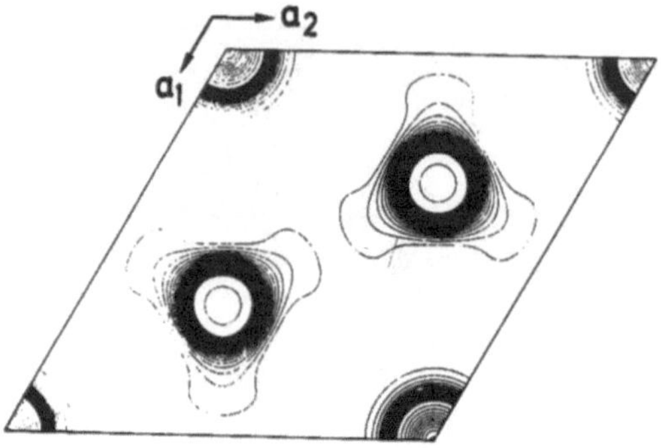

Fig. 12. Probability density map. Gram-Charlier expansion up to the 6th order.

Fig. 16 is not located at the "atomic position" of the Ag ion, but at a ringlike region around this position. Just this result shows how powerful these structure refinements are and how important it is to transform the refined coefficients into a pdf. It should be noted, that the Ag density (Fig. 20) shows a maximum at (1/2,1/2,0), but this is only caused by the overlap of the electron clouds of the silver ions and it is not related to the residence probability of the Ag ions. Similar structure investigations (except the calculations of pdf-maps) by Cava, Reidinger and Wuensch [21] have resulted in the Ag-ion arrangement in α-AgI as shown in Fig. 2. From this recent structure investigation it is concluded that the Ag-ion occupies only the tetrahedral sites and that they move through the lattice along the heavy solid lines in Fig. 2. They jump from tetrahedron to tetrahedron by passing through the common face of these tetrahedra. It follows, that there exists no difference in the Ag-coordination between the α- and ß-form of AgI. In ß-AgI these tetrahedra are connected by common edges, whereas in α-AgI they are connected by common faces. The latter arrangement offers a diffusion path with very low potential differences. From this point of view the arrangement of th I_4 tetrahedra is the main reason for the high degree for occupational disorder in the α-form.

Crystal Structure and Crystal Field Potential

The density calculations and the analysis of the anharmonic thermal vibrations reveal the importance of potential considerations. High ionic conduction requires nets of three-dimensional, two-dimensional or one-dimensional paths with low potential differences. Jumps between occupied and unoccupied sites require defects. However, the deviations from a complete order may be so small that it is difficult to detect them with diffraction methods as it is shown for lithium nitride.

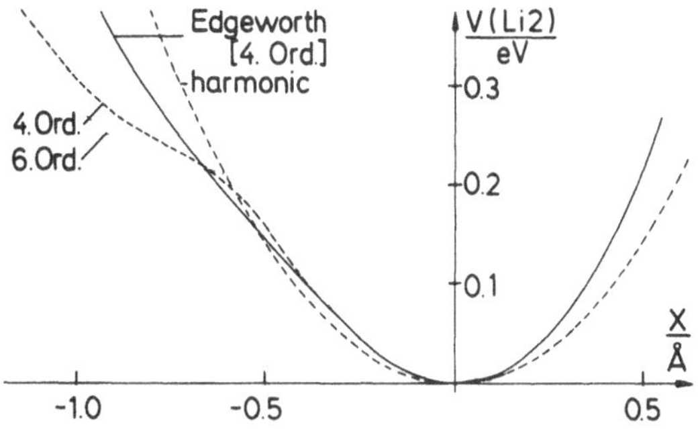

Fig. 13. Li(2) potential along the line N-Li(2)-Li(2)-N (cf. Fig. 6) for different anharmonic temperature factors. The curves based on the Gram-Charlier Formalism are only designated by the order.

Potential calculations for ionic conductors have been already carried out for several compounds, e.g. AgI, Fluorites and Li_3N. [10 24 25] Also molecular dynamics have been applied to the ionic conductor CaF_2 [26]. These calculations are based on assumption on Coulomb interactions, polarizations and repulsion forces. Usually these assumptions are open to some doubt and the results need confirmation by experimental methods.

One way for the experimental determination of an atomic potential is based on the probability density maps [20]. These maps allow in a straight forward way the calculation of the corresponding potential. This can be shown in the following way: The temperature factor is determined by the time averaged mean value of the exponential function of the atomic displacements:

$$T_j(\underset{\sim}{h}) = \langle \exp(2\pi i \, \underset{\sim}{h} \, \Delta\underset{\sim}{x}_j) \rangle_t$$

The displacements $\Delta\underset{\sim}{x}_j$ have to be weighted by the Boltzmann factor. Eq. (8) can then be written as:

$$T_j(\underset{\sim}{h}) = \frac{\exp(2\pi i \, \underset{\sim}{h}\underset{\sim}{x}_j) \exp(-V_j(\underset{\sim}{x}_j)/KT) \, d^3 x_j}{\exp(-V_j(\underset{\sim}{x}_j)/KT \, d^3 x_j} \quad (9)$$

K: Boltzmann constant

V_j: Potential of the j-th atom

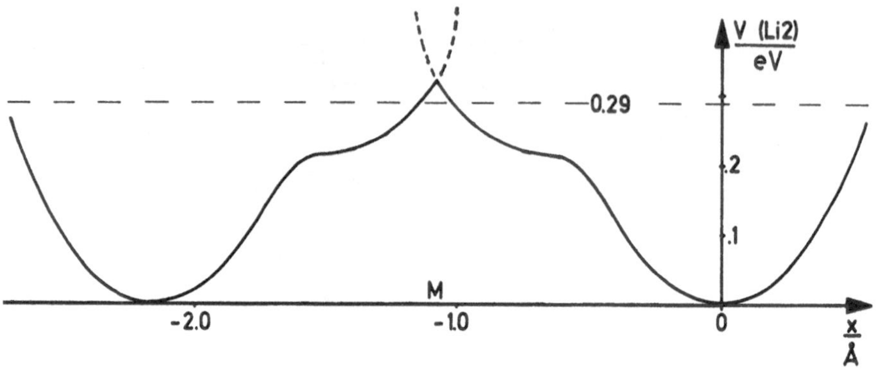

Fig. 14. Estimation of the potential between an occupied and an unoccupied Li(2) site.

The nominator of Eq. (9) represents a Fourier transform. As shown above, T_j is also the Fourier transform of the pdf. Therefore holds

$$pdf_j(\underset{\sim}{x}) \quad = \quad \exp(-V_j(\underset{\sim}{x})/KT)/N_j$$

N is the denominator of Eq. (9). $V_j(x)$ represents an energy. Its zeropoint can be chosen arbitrarily, e.g. in the following way:

$$V_j(\underset{\sim}{x}_o) \quad = \quad 0$$

$\underset{\sim}{x}_o$: average position of the atom j.

It follows

$$\ln N_j \quad = - \ln(pdf(\underset{\sim}{x}_o))$$

$$V_j(\underset{\sim}{x}) \quad = -KT(\ln(pdf(\underset{\sim}{x})-\ln(pdf(\underset{\sim}{x}_o)) \tag{11}$$

Eq. (11) shows the potential of an atom can be calculated for each symmetry and for any direction, if a probability density function is available. This potential can be derived from elastic diffraction experiments.

This shall now be demonstrated for Li_3N. The Li(2) potential of Fig. 13 is derived from the pdf of Fig. 12. It is the potential along the long diagonal of Fig. 12, this means along the line N-Li(2)-Li(2)-N. It demonstrates clearly that the potential is strongly asymmetric. It shows further the potentials derived from structure refinements with anharmonic temperature factors up to fourth

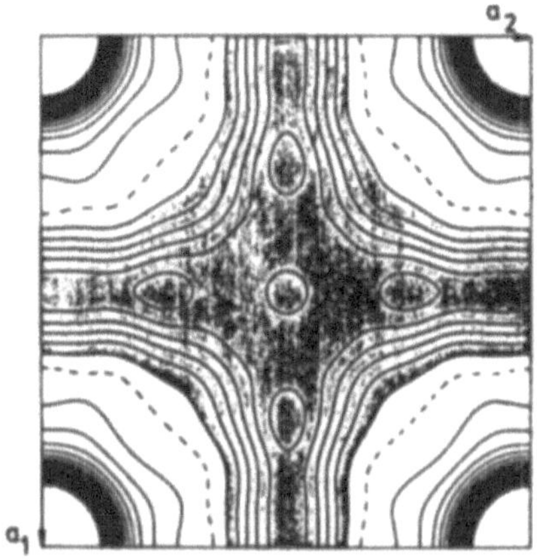

Fig. 15. Partial Ag-electron density of α-Ag_3SI in the (100) plane.

order for the Gram-Charlier and for the Edgeworth expansion and for refinements with harmonic temperature factors. It is evident, how insufficient the Edgeworth and the harmonic refinements approximate the real potential.

An estimation of the potential between two Li(2) positions is shown in Fig. 22. It was generated by mirrowing Fig. 13 at the mirrow plane, which coincides with the small diagonal of Fig. 18. The potential at the mirrow point M has a height of about 0.30eV. This compares very well with the activation energy of 0.29eV, which has been found in conductivity measurements[11]. In this way it seems to be possible to estimate the activation energy of the mobile ions from elastic diffraction experiments. It should be taken in mind, that in the construction of Fig. 14 the repulsion terms are not drawn, which must be present, if both Li(2) positions are occupied. Fig. 14 should be considered as an assessment of the potential, if one Li(2) position is vacant.

In Ag_3SI the Ag ions diffuse along the symmetrically equivalent directions (1/2,x,0,x =0-1) through the lattice (Fig. 15). Their potential along this path shows Fig. 17. It was calculated from the pdf-maps of Fig. 16 in the above described way. The potentials for the three "atomic positions" overlap. It is now assumed, that only those parts of these potentials are relevant for the diffusion of the Ag-ions, which are drawn by heavy black lines in Fig. 14. The highest potential barrier has then a height of about 0.03 eV, which compares well with 0.04 eV from conductivity measurements[27,28]. The examples of Li_3N and Ag_3SI show that one of the best ways to analyse the conduction mechanism in a special crystal is the interpretation of diffraction data. This can be done in the form of density maps, anharmonic temperature factors,

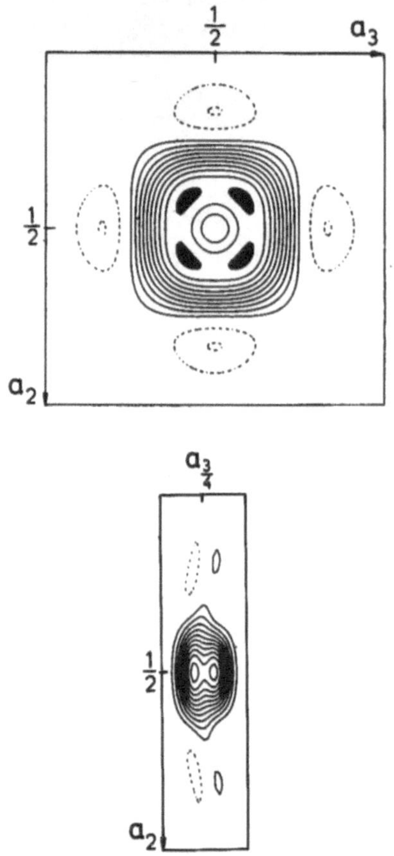

Fig. 16. Ag-Probability density maps in α-Ag_3SI Gram-Charlier Expansion up to the 6. order
(a) Atomic position (0,1/2,1/2) and symmetrically equivalent positions
(b) Atomic position (1/2,1/4,0) and symmetrically equivalent positions.

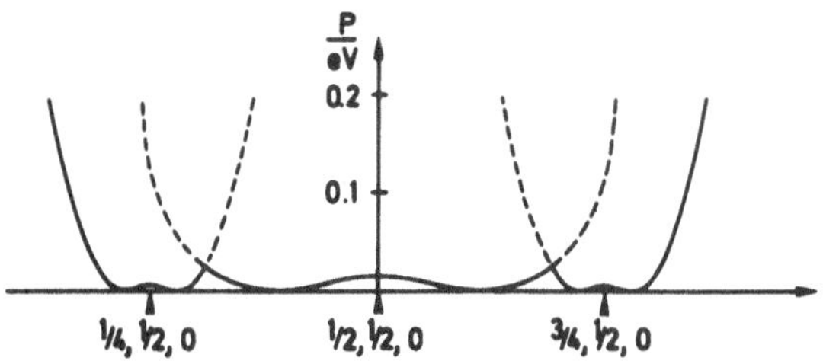

Fig. 17. Estimated potential along the diffusion path of the Ag ions in α-Ag_3SI.

probability density-maps and atomic potentials. The crystal field potentials of fast ionic conductors have to have low differences along the conducting paths and the regular lattice sites must have local but weak potential minima within such paths.

References

1. M. Faraday, "Experimental Researches in Electricity", Taylor and Francis, London (1839).
2. E. Warburg, Wiedemann Ann. Phys. 21, 622 (1884).
3. W. Nernst, Z. Elektrochem. 6, 41 (1899).
4. C. Tubandt and E.Z. Lorenz, Physik. Chemie 87: 513 and 543 (1914).
5. L.W. Strock, Z. Physik. Chemie B 25: 411(1934)and B 31: 132 (1936).
6. P. Hagenmuller and W. van Gool (Eds.), "Solid Electrolytes", Academic Press, New York, 1978.
7. P. Vashishta, J.N. Mundy and G.K. Shenoy (Eds.), "Fast Ion Transport in Solids", Elsevier North Holland, New York, 1979.
8. R.A. Huggins, in A.R. Cooper and A.H. Heuer (Eds.) "Mass Transport Phenomena in Ceramics", Plenum, New York, 1975, p. 155.
9. R.H. Radzilowski and J.T. Kummer, J. Electrochemic. Soc. 118: 714 (1971).
10. W.F. Flygare and R.A. Huggins, J. Phys. Chem. Solids 34: 1199 (1973).
11. U. v. Alpen, A. Rabenau and G.H. Talat, Appl. Physics Lett. 30: 621 (1977).
12. A. Rabenau in: Festkörperprobleme (Advances in Solid State Physics) Volume XVIII, J. Treusch (Ed.) Vieweg, Braunschweig, 1978, p. 77.

13. A. Rabenau and H. Schulz, J. Less-Com. Metals 50: 155 (1976).
14. H. Schulz and K.H. Thiemann, Acta Cryst. A 35: 309 (1979).
15. H. Schulz and U. Zucker, in: "Fast Ion Transport in Solids" (Eds. P.Vashishta, J.N. Mundy and G.K. Shenoy), Elsevier North Holland, New York, 1979, p. 495.
16. F.Y Edgeworth, Trans. Camb. Phil. Soc. 20: 35 (1905).
17. M.G. Kendall and A. Stuart, "The advanced Theory of Statistics" Vol. I, Giffin, London, 1963.
18. "International Tables for Crystallography" Vol. IV, Kynoch Press, Birmingham (1974).
19. C.K. Johnson, Acta Cryst. A 25: 187 (1969).
20. U. Zucker, Thesis, University Karlsruhe, Germany, (1980).
21. R.J. Cava, F. Reidinger and B.J. Wuensch, Solid State Com. 24: 411 (1977).
22. R.J. Cava, F. Reidinger and B.J. Wuensch, J. Solid State Chem. 31: 69 (1980).
23. E. Perenthaler and H. Schulz, Acta Cryst. A (1981) in the press.
24. O.B. Ajayi, "Calculation of motional activation enthalpy in fast ionic conductors", Thesis, Stanford University (1975).
25. J.R. Walker and C.R.A. Catlow, in: "Fast Ion Transport in Solids" (Eds. P. Vashista, J.N. Mundy and G.K. Shenoy), Elsevier, North Holland, New York, 1979), p. 491.
26. G. Jacucci and A.J. Rahman, Chem. Phys. 69: 4117 (1978).
27. B. Reuter and K. Hardel, Ber. Bunsenges. Physik. Chem. 70: 70 (1966).
28. G. Chiodelli, A. Magistris and A. Schiraldi, Z.f. Physikal. Chem. Neue Folge 118: 177 (1979).

THERMODYNAMIC AND TRANSPORT PROPERTIES OF SUPERIONIC CONDUCTORS AND ELECTRODE MATERIALS

J.H. Harding

Theoretical Physics Division
AERE Harwell
Didcot, Oxon

INTRODUCTION

The terms "superionic conductor" and "solid electrolyte" are used to denote those solids that exhibit anomalously large ($\sim$ 1 (ohm-cm)$^{-1}$) ionic conductivities. Such a definition is rather too wide since most solids have high ionic conductivities provided the temperature is high enough so it is common to suggest further features of this class of material. These include

(i) weak temperature dependence of the ionic conductivity

(ii) low activation energies (few tenths of an eV)

(iii) anomalously low prefactors in the Arrhenius expression.

Exceptions can be found to all of these criteria; for example stabilised zirconia has an activation energy of about 1 eV and yet is counted a superionic material in most of the literature.

Superionic materials undergo a wide range of phase transition behaviour, as indeed do many solids; and as is well known some phases are superionic and some are not. The most well known case is AgI where the Wurtzite-bcc first order transition is accompanied by a change in conductivity of four orders of magnitude. However it need not always be the high-temperature phase that is superionic. In CuBr and Ag_3SI both high and low temperature phases are, although there is a discontinuity in the conductivity at the boundary and in the case of Na_2WO_4 and Na_2MoO_4 the superionic phase exists at intermediate temperatures, both low and high temperature phases being "normal" (Broers, 1958, Bottelberghs, 1976).

Superionic conductors obey the basic classic equations of conductivity; they are unusual in having parameters that are 'optimised for high conductivity'. In the rest of this lecture we shall consider to what extent a hopping model may be used to describe these materials, and what adjustments may be necessary when the simple model breaks down.

IONIC THEORY

In general we may describe the conductivity of a material by the equation

$$\sigma = \sum_i n_i \mu_i (z_i e) + n\mu_e \rho$$

where n_i, μ_i, z_i are the number per unit volume, mobility and charge of the ionic species present and the second term is an electronic contribution. In most cases of current interest this can be neglected and further, ionic transport confined to one sub-lattice. If we can assume that charge and mass are transported by the same mobile species and that correlation effects may be neglected, the mobility μ_j is related to the diffusion constant D_j by the Nernst-Einstein relation

$$\frac{\mu_j}{D_j} = \frac{z_j e}{kT}$$

This can be extended, for self diffusion and isotope diffusion, to cover cases where correlation is important by multiplying the denominator of the right-hand side by a correlation factor f. The problem of calculating this will be briefly considered later.

The diffusion coefficient D_j need not be the same as that measured by tracer experiments D_{Tj}. The ratio between these $H_R = D_j/D_{Tj}$, often referred to as the Haven ratio, has frequently been used to identify the transport mode. Murch (1979) has shown further that this ratio can give a great deal of information about correlation processes since, if we define the tracer correlation factor, f_T, then in cases where the transport mode does not affect H_R (i.e. no neutral defects and no electrical conductivity)

$$H_R = \frac{f_T}{f}$$

f_T contains, in addition to physical correlation, a configurational term due to the tagging of the tracer atom and so H_R is the ratio between the walks of marked and unmarked atoms. Murch shows that, for an athermal lattice gas in a simple cubic system, H_R has a minimum near a second-order phase transition, a point that is of some interest in considering fluorites.

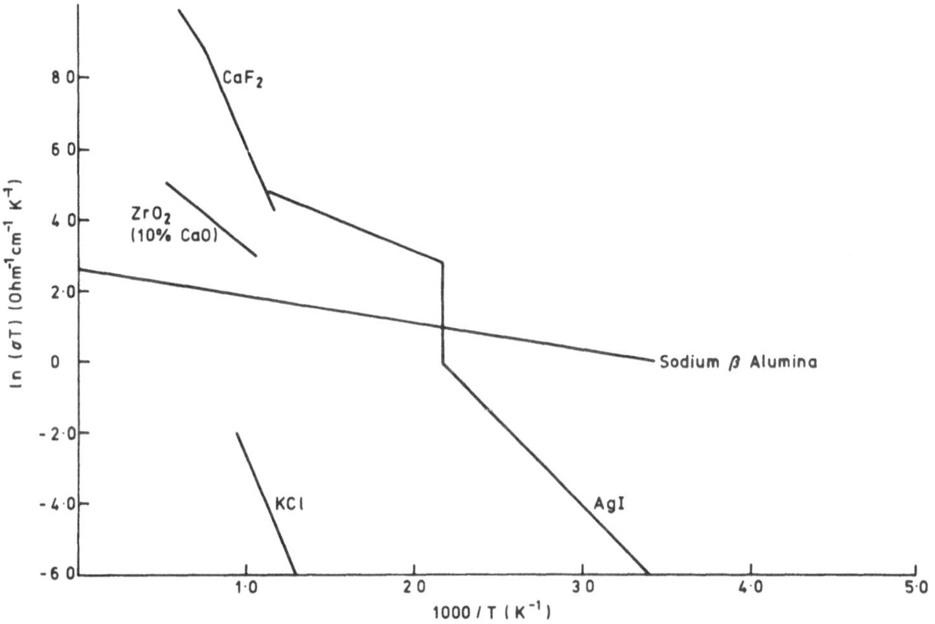

Fig. 1. Conductivities of some superionic conductors.

For an ideal lattice, in the case of the completely random walk, with independent mobile ions, we may write the mobility as

$$\mu_i^{(\lambda)} = \frac{z_i e}{2kT} C_i \sum_\alpha \Gamma_\alpha \chi_\alpha^2$$

where C_i is the mole fraction of defects that are responsible for the conduction and Γ_α is the jump frequency for jumps of type α having a projected displacement of magnitude χ_α along the λ axis. If more than one ion is involved in each defect jump we must also multiply by the number of ions involved, m_i. If we assume that the lattice is cubic and that only one jump type involving only one ion is possible (e.g. the simple vacancy mechanism) the expression for the conductivity takes the simple form

$$\sigma = \frac{nc}{6kT} (ze)^2 \Gamma r^2$$

where r is the jump distance.

In the case of simple defect conduction, c may be obtained from the mass action law. This has the effect of making the observed activation energy dependent on the enthalpy of formation, ΔH_F, of the defect. This can be calculated by HADES-type programs. Although these calculate an energy at constant volume, and what is required is an enthalpy at constant pressure this does not in fact affect the result under normal conditions, since to first order in the defect concentration $\Delta G_p \sim \Delta A_v$ (Lidiard et al., 1980). The value of this ΔH_F term may be several volts; and so, from the observed activation energies of superionics it is clear that the density of defects must be determined by other considerations. Quite what these are is a matter of continuing debate and it is most unlikely that the answer is the same for all superionics.

Some phenomenological models have suggested that the defect concentration may be as high as 50% - the so-called sub-lattice melting argument. However, more recent experimental evidence, supported by some molecular dynamics calculations (Dixon and Gillan 1980) suggests that the defect concentration may only be a few per-cent, more in keeping with ΔH_F values suggested by static-lattice calculations (Catlow 1980).

CALCULATION OF THE JUMP FREQUENCY, Γ

A hopping model assumes that the ions spend most of their time in random vibrations about a lattice site, from time to time jumping to a neighbouring site. The calculation of the jump frequency, Γ, is thus at the heart of such a model. It should also be clear that in the case of superionics such a model is an idealisation and perhaps a risky one. If we consider the case of AgI at 150°C, the time of flight of an ion is of the order of 7×10^{-13} sec. However the random-hopping model requires that τ, the time between hops $(1/\Gamma)$ be about 5×10^{-12} sec. to explain the conductivity. This difference of only an order of magnitude is also seen in the fluorites as shown in the molecular dynamics calculations referenced above. The effect of a breakdown in the model will be two-fold. The effective activation energy will usually rise because of defect association, and the pre-exponential factor will fall because of blocking effects. Thus the hopping model may be expected to give some kind of upper bound to the conductivity. The effect of 'sub-lattice melting' if such a thing occurs is to make conductivity less efficient, an effect shown by the dramatic fall in conductivity of AgI when the lattice melts at 555°C.

This calculation of the jump frequency, Γ, was treated by Vineyard (1957) who showed that for a quasi-harmonic system a general expression for the jump frequency could be obtained.

Consider a vacancy in a crystal and let Γ be the average rate at which a specific adjacent atom jumps into the vacant site. Let the crystal contain N atoms (i.e. 3N degrees of freedom $X_1 \ldots\ldots X_{3N}$) and for convenience let us define a set of mass weighted coordinates $Y_i = X_i \sqrt{m_i}$. There will be a configuration for which the potential function $\Phi(Y_1 \ldots\ldots Y_{3N})$ is a minimum and a similar minimum corresponding to having the vacancy in a neighbouring position. There must be at least one saddle point between these two minima (A and B in figure 2) and for most cases of interest there will only be one (P) on the N-1 dimensional hypersurface in the diagram. There exists a unique hypersurface S, passing through P and orthogonal to the contours of constant Φ everywhere else. This separates A from B and any point from the area surrounding A reaching S with finite velocity will go into B. At thermal equilibrium there are a number of representative points, Q_A, in the region to the left of S, a number, I, crossing S from left to right per second. Hence

$$\Gamma = \frac{I}{Q_A}$$

For classical systems in equilibrium the position and velocity of a representative point in configuration space are independently distributed and the density, ρ may be written as

$$\rho = \rho_o \exp(-\Phi/kT)$$

From this we can obtain an expression for Γ which is

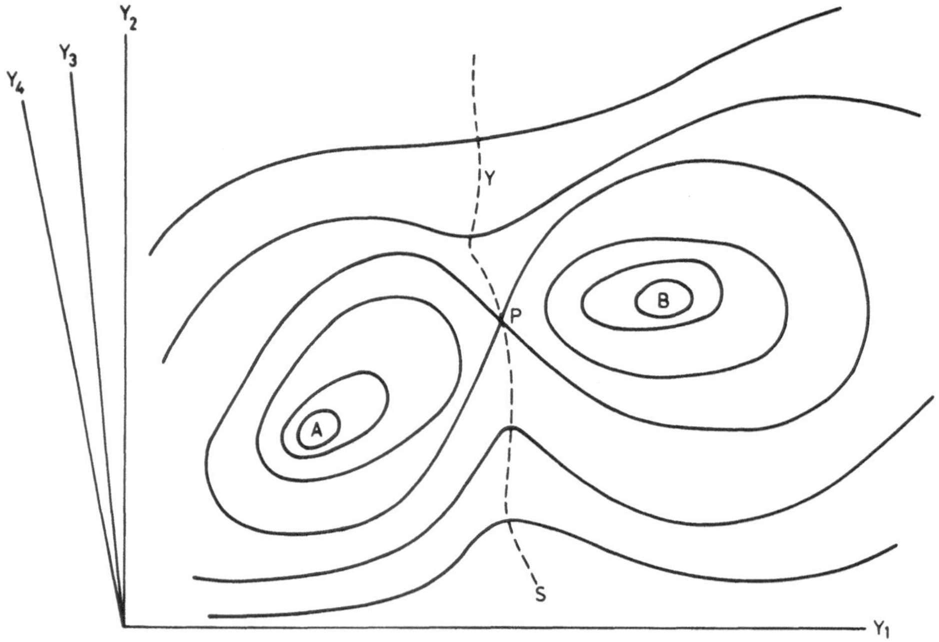

Fig. 2. Schematic potential surface for simple vacancy mechanism (after Vineyard (1957)).

$$\Gamma = \sqrt{\frac{kT}{2\pi}}\,\frac{\int_S \exp(-\frac{\Phi}{kT})\,dS}{\int_A \exp(-\frac{\Phi}{kT})\,dv.}$$

We may apply the theory of small vibrations to this and expand Φ near the points A and P in a Taylor series to second order in the normal coordinates q_j and frequencies ν_j

$$\Phi = \Phi(A) + \frac{1}{2}\sum_{j11}^{N} (2\pi\nu_j)^2\, q_j^2$$

and so obtain our final expression for Γ.

$$\Gamma = \frac{\prod_{j=1}^{N} \nu_j}{\prod_{j=1}^{N-1} \nu_j'} \exp(-\{\Phi(P) - \Phi(A)\}/kT)$$

Since $\Phi(P)$, $\Phi(A)$ represent points of potential energy the migration energy $E_m = \Phi(P)-\Phi(A)$ may be obtained from a static lattice calculation, either a simulation like HADES or, as has been tried (McOmber et al., 1979), an electronic structure calculation. There may be practical difficulties in low symmetry systems in deciding on the position of the saddle point but these can usually be overcome by doing a set of calculations to map out the shape of the energy surface in the region of interest.

The other term in our expression has the dimensions of a frequency and is often taken to be the Einstein or Debye frequency of the jumping atom. This however is rather crude since in general the frequencies will be quite strongly perturbed by the motion of the ions in the lattice. The resemblance of this term to an entropy term (minus a degree of freedom) has often been noted and so this is frequently referred to as an entropy of activation. The calculation of entropies is in itself a matter of interest and to this we now turn.

CALCULATION OF ENTROPIES

This has in the past posed considerable problems; however we have recently (Harding and Stoneham, 1981) reported a method by which reasonable estimates for these quantities may be obtained without the necessity of long and tedious calculation.

We consider a group of ions comprising the defect (or saddle-point) and nearest neighbours. In general it is found that it is sufficient to consider the first and second nearest neighbour only; the problems in considering more are practical rather than theoretical. The positions of these cluster ions are calculated using the HADES routines and so explicit account is taken of local lattice distortion. We then impose cyclic boundary conditions on the problem. In order to do this in a consistent manner it is generally necessary to embed the cluster in a region of perfect crystal. The result of this procedure is to define a unit cell for the distorted lattice with the same bulk volume as a number of primitive unit cells in the perfect lattice. Physically this amounts to embedding the defect cluster in a regular array of defects. We then obtain the frequency spectrum for this periodic structure using a shell model calculation.

If we wish to obtain a true entropy from this we must obtain the partition function, Q, which in the quasi-harmonic approximation may be written as

$$\ell nQ = - \int_0^\infty d\underline{k} \sum_n \ln(1 - \exp(\frac{h\nu_{nk}}{kT}))$$

The effect of using a non-primitive unit cell is to allow us to replace the integral over $\underline{k}$ by an implicit summation over the important parts of $\underline{k}$ space. In the case of a perfect crystal the Brillouin zone of a large cell constructed from N primitive cells has a volume of 1/N of the primitive cell Brillouin zone; thus the zone bands are bent back onto the $\underline{k} = 0$ vector and so taking the $\underline{k} = 0$ vector for this large cell is equivalent to taking a properly weighted sum over the zone for the primitive cell provided the large cell is correctly chosen. Procedures for choosing a sensible cell have been obtained (Evarestov, 1975); for example a cell in the fcc lattice with lattice vectors {220} allows us to sum over the Γ X and L points in the primitive zone. In the defect crystal the same method picks up the perturbed phonon spectrum corresponding to this summation. We can thus obtain an entropy change (due to the vibrational terms) for the process. There may of course be configurational terms but these can be obtained by standard methods. It should be noted that the result is an entropy at constant volume. To obtain an entropy at constant pressure (the usual experimental situation) we must add a correction (Lidiard et al., 1980)

$$\Delta Sp = \Delta Sv + \Delta Vp \frac{\alpha_p}{K_T}$$

where α_p is the thermal expansion coefficient and K_T is the bulk modulus. To get an entropy of activation (which is ΔS_V) we must remove the mode that is acting as the reaction coordinate, ν^*.

If we have found the true saddle point this will have a negative eigenvalue and also its eigenvector will be heavily weighted in the direction of the reaction coordinate. Thus we may calculate $\tilde{\nu}$, the effective frequency in the Vineyard model as

$$\tilde{\nu} = \nu^* \exp\left(\frac{\Delta S^\ddagger}{k}\right)$$

Before we turn to the results of specific calculations, I would like to consider, in outline at least, the correlation factor f mentioned earlier.

THE CORRELATION FACTOR

The classic review of this area is that of LeClaire (1970). It is convenient to begin with a rather closer analysis of the diffusion equation itself. Let us consider a system where there exist species whose migration can be followed. We shall consider chemically homogeneous systems so that the relative probability that after time t an atom has migrated distance X measured along the n axis, f(X,t) is independent of position. Let the species concentration be c(x) at x at t=0. Then the number of atoms on the left of a plane at x_0 perpendicular to the x axis which will be found on the right of it after time t is

$$\int_{\infty}^{\infty} c(x) \left\{ \int_{x_0 - x}^{\infty} f(X,t)\,dX \quad dx \right\}$$

and a similar expression obtains for those moving the other way. Since f(X,t) falls off rapidly with X, we can expand c(x) about $c(x_0)$ and so an equation may be obtained for J_x, the difference between the two integrals as

$$J_x = - \frac{\langle X^2 \rangle}{2t} \frac{dc}{dx}$$

from which D_x is obtained by comparison with Ficks law.

If we consider $x_1, x_2 \ldots x_n$ as the components of n jumps of an atom along the x axis made, on average, in time t

$$\langle X \rangle^2 = \langle \sum_{i=1}^{n} x_i^2 \rangle$$

$$= \sum_{i=1}^{n} \langle x_i^2 \rangle + 2 \sum_{i=1}^{n-1} \sum_{j=1}^{n-i} \langle x_i x_{i+j} \rangle$$

The second term is that describing the correlations; this equation is usually expressed in the form

$$\langle X^2 \rangle = f \sum_{i=1}^{n} x_i^2$$

where f, the correlation factor is

$$f = 1 + \frac{2 \sum_{i=1}^{n-1} \sum_{j=1}^{n-i} \langle x_i x_{i+j} \rangle}{\sum_{i=1}^{n} \langle x_i^2 \rangle}$$

In simple cases this can be obtained analytically; for example the correlation factors for self-diffusion by single vacancies for certain lattices are pure numbers

Lattice	f
Diamond	0.5
Simple cubic	0.65311
bcc	0.72722
fcc	0.78146
hcp	$f_x = f_y = 0.78121$: $f_z = 0.78146$

In more complex systems, where the correlation factor is a function of position and jump frequency, various methods of calculation are available, for example Monte Carlo as suggested by Murch and Thorn (1979).

We have now assembled the various pieces we will require for a hopping model, and it is thus appropriate to consider some experimental systems.

CALCULATIONS ON FLUORITE STRUCTURES

From a theoretical point of view the fluorites (or antifluorites) are a convenient set of structures to choose. They are often ionic and so simple shell-model potentials are appropriate; they also have a simple structure (all ions on cubic sites) that makes analysis of possible migration pathways comparitively simple and cheap compared to less symmetric systems. A considerable amount of work has already been done, summarised by Catlow in a recent comment (1979). Here I propose to discuss the alkaline earth fluorides briefly, and suggest extensions of these calculations to other systems.

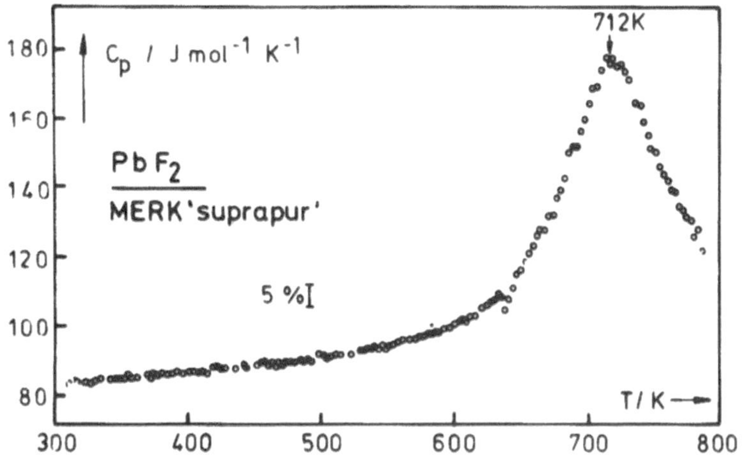

Fig. 3. Molar heat capacity of PbF_2 (Nolting, 1980).

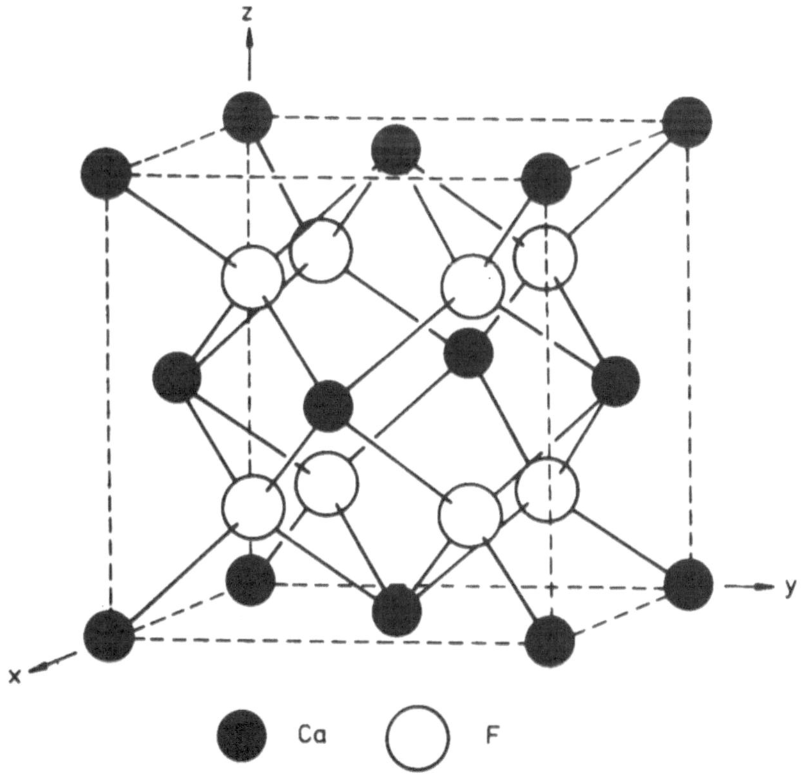

Fig. 4. Clinographic projection of the unit cell of the cubic structure of fluorite, CaF_2.

The fluorite and antifluorite structures undergo a reordering at about 80% of their melting point to produce a superionic system. This is accompanied by a specific heat anomaly, the best-characterised example is that of PbF_2 (Figure 3). This may be a diffuse second-order transition; the data is not decisive. Typical values for the transition point (T_c) and the melting point (T_m) are shown in the Table.

Compound	T_c(K)	T_m(K)
CaF_2	1423	1691
BaF_2	1233	1593
PbF_2	703	1095
K_2S	1050	1221
Na_2S	1273	1442

Good experimental data for diffusion are available for PbF_2, BaF_2 and $SrCl_2$ (Carr et al. 1978, Figueroa et al. 1978); the latest data available on CaF_2 is that of Derrington et al. (1975) and was obtained from an experimental setup that was not suitable for high-temperature work (cf.Carr). Data on Na_2S, K_2S is sufficient to establish the existance of the superionic phase but no more (Dworkin and Bredig, 1968, Möbus et al. 1964).

Defect calculations have been performed by Catlow and Norgett (1973) and are shown in the following Table. Various diffusion mechanisms are possible (some examples are shown in fig. 5); the lowest energy one is the vacancy mechanism.

	Frenkel Energy (eV)	defect migration energy (eV) (vacancy mechanism)
CaF_2	2.67	0.27
BaF_2	1.91	0.39

Using the methods outlined above we may obtain the pre-exponential factors.

	ν^*(Hz)	$\Delta S^{\pm}$	$\Delta S_{Frenkel}$	σT(theo)(Ωcm)$^{-1}$	σT(expt)(Ωcm)$^{-}$
CaF_2	3.75×10^{12}	-3.4k	3.2k	5.7×10^{8}	-
BaF_2	3.57×10^{12}	0.27k	2.4k	1.5×10^{9}	1.57×10^{8}

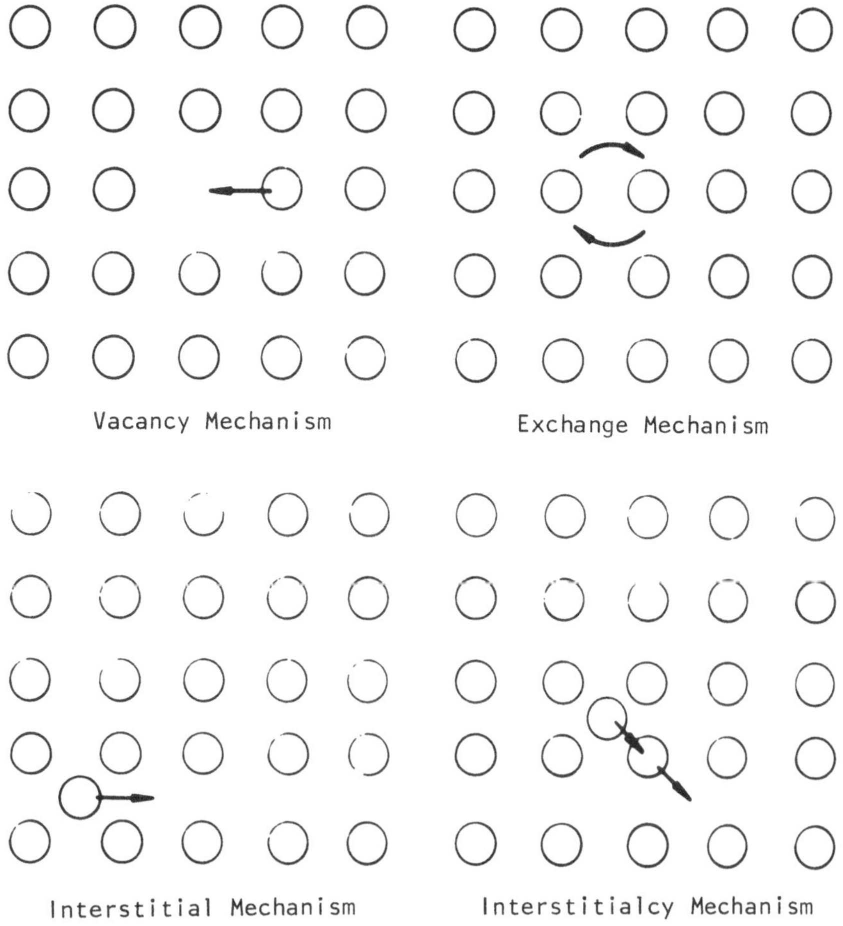

Fig. 5. Various defect mechanisms for a simple cubic lattice.

This shows that the calculated values of ν^* are considerably lower than the Debye frequency often used. The results here agree to within about an order of magnitude which considering the approximations made must be considered fair.

CONCLUSIONS

We have considered the hopping model as a possible description of superionic conductors. It is clear that it is an idealisation; in many cases it may be adequate to consider groups of associated defects as a method of correcting for high defect concentrations but beyond this it seems necessary to move into some lattice gas type of model. The advantage however, of the hopping model is that all the parameters it requires are calculable or can, in the case of the defect concentration, be estimated from experiment. This gives it a considerable advantage over the parameterised thermodynamic models in that we have a much clearer ability to test the assumptions built into the calculation against experiments.

REFERENCES

Broers G.H. (1958) PhD Thesis (Amsterdam).
Bottelberghs P.H. (1976) PhD dissertation (Utrecht).
Carr V.M., Chadwick A.V., Saghafian R. (1978) J.Phys. C11 L637.
Catlow C.R.A., Norgett M.J. (1973) J.Phys. C6 1325.
Catlow C.R.A., Diller K.M., Norgett M.J. (1977) J.Phys. C10 1395.
Catlow C.R.A. (1980) Comments Solid State Phys. 9 157.
Derrington C.E., Navrotsky A., O'Keefe M. (1976) Sol.St.Comm. 18 47.
Dworkin A.S., Bredig M.A. (1968), J.Phys.Chem. 72 1277.
Evarestov R.A. (1975) Phys.Stat.Sol. (b) 72 569.
Figueroa D.R., Chadwick A.V., Strange J.H. (1978) J.Phys. C11 55.
Gillan M.J., Dixon M. (1980) J.Phys. C13 1901, 1919.
Harding J.H., Stoneham A.M. (1981) Phil.Mag. in press.
LeClaire A.D. (1970) in Physical Chemistry (an advanced treatise) ed. W. Jost Publ. Academic Press, New York.
Lidiard A.B., Catlow C.R.A., Cornish J., Jacobs P.W.M. (1980) to be published.
McOmber J.I., Topiol S., Ratner M.A., Shriver D.F. (1979) J.Phys. Chem. Solids 41 447.
Möbus H.H., Witzmann, H., Harting R. (1964) Z.Phys.Chem. 40 227.
Murch G.E. (1979) Phil.Mag. A41 701.
Murch G.E., Thorn R.J. (1979) Phil.Mag. A39 673.
Nölting J. (1980) to be published, (quoted in C.R.A. Catlow (1980) Comments Solid State Phys. 9 157).
Vineyard G.H. (1957) J.Phys.Chem.Sol. 3 121.

EVALUATION AND MEANING OF IONIC AND DIPOLAR LATTICE SUMS

Edgar R. Smith[a] and John W. Perram[b]

[a]Department of Mathematics
University of Melbourne
Parkville, Vic. 3052
Australia

and

[b]Matematisk Institut
Odense Universitet
Campusvej 55
5230 Odense M
Denmark

ABSTRACT

We discuss the lattice sums which arise in simulations of ionic and dipolar systems using periodic boundary conditions. The lattice sums are conditionally but not absolutely convergent and so the order of summation must be given. We sum them by spherical shells and discuss the role of an "external dielectric constant" ε', the dielectric constant of a continuum outside the outermost shell becomes large. The lattice sums are converted into absolutely and rapidly convergent lattice sums and methods for rapid evaluation of these sums are discussed. The effect of ε' on the Hamiltonian is discussed. The effect of changes in ε' on the mean square dipole moment fluctuations is discussed and illustrated by reference to some Monte Carlo simulations of restricted primitive electrolyte.

I. INTRODUCTION

Molecular dynamics and Monte Carlo simulations of bulk systems are almost always carried out in the minimum image (MI) convention, but other types of boundary condition on the simulation sample may be considered a part. There are two widely used methods: the reaction field method of Barker and Watts [1] (RF) and periodic boundary conditions (PBC). We must first consider the Hamiltonian which is produced by these assumptions and then discuss the way that these Hamiltonians affect the results obtained.

We begin with some definitions of the system. We have a set of N particles in a cubic box of side L, at positions $\underset{\sim}{r}_1,\ldots,\underset{\sim}{r}_N$ and with interactions which are composed of short-ranged interactions and long-ranged interactions. The short-ranged interactions may be written in the form $\phi_{ij}{}^S(\underset{\sim}{r}_{ij})$, and for this potential to be "short ranged" we require that

$$|\phi_{ij}{}^S(\underset{\sim}{r}_{ij})| < D|\underset{\sim}{r}_{ij}|^{-3-\varepsilon} \tag{1.1}$$

for $\varepsilon > 0$. The long-ranged interactions are either those between charges

$$\phi_{ij}{}^C(\underset{\sim}{r}_{ij}) = q_i q_j |\underset{\sim}{r}_{ij}|^{-1} \tag{1.2}$$

or dipoles

$$\phi_{ij}{}^D(\underset{\sim}{r}_{ij}) = \underset{\sim}{\mu}_i \cdot \left[I - \frac{3\underset{\sim}{r}_{ij}\underset{\sim}{r}_{ij}}{|\underset{\sim}{r}_{ij}|^2}\right] \cdot \underset{\sim}{\mu}_j/|\underset{\sim}{r}_{ij}|^3 \tag{1.3}$$

embedded in the particles. Charge-dipole interactions are also long-ranged, but we do not consider them here. To use MI we first construct a periodic array of replicas to the simulation cube which fill space. Thus for each particle i at $\underset{\sim}{r}_i$ in the original simulation cell, there is an infinite set of replicas of the particle at the points $\underset{\sim}{r}_i + L\underset{\sim}{n}$ where $\underset{\sim}{n} = (n_1,n_2,n_3)$ is a vector with integer components. We define $\underset{\sim}{r}^*_{ij}$, the minimum image $\underset{\sim}{r}_{ij}$ as that vector $\underset{\sim}{r}_i - (\underset{\sim}{r}_j + L\underset{\sim}{n})$ which has minimum length, the minimization being over all $\underset{\sim}{n}$. In MI, a particle i at $\underset{\sim}{r}_i$ in the simulation cell interacts not necessarily with a particle j at $\underset{\sim}{r}_j$ but with its nearest replica, it being assumed that the set of replicas includes the original particle. Thus in MI, the Hamiltonian takes the form

$$H_{MI}(\underset{\sim}{r}_1,\ldots,\underset{\sim}{r}_N) = \frac{1}{2}\sum_{i=1}^{N}\sum_{\substack{j=1\\ i\neq j}}^{N}(\phi_{ij}{}^S(\underset{\sim}{r}_{ij}{}^*) + \phi_{ij}{}^L(\underset{\sim}{r}_{ij}{}^*)) \tag{1.4}$$

where $\phi_{ij}{}^L(\underset{\sim}{r})$ is the charge (eq. (1.2)) or dipole (eq. (1.3)) interaction.

In reaction field boundary conditions, the energy of a given particle is estimated as follows. First, draw a sphere of radius R (normally $R \leq L/2$) about the particle, and include all the MI interactions with particles inside the sphere. Then imagine that the sphere is surrounded by a dielectric continuum of dielectric constant ε'. These boundary conditions are not normally used for ionic systems, so we shall concentrate on the dipole case. The particles will have a net dipole moment which will polarize the continuum, setting up a reaction field which interacts with the dipoles in the sphere, and half of this energy is associated with the dipoles and half with the continuum. This method is discussed in the next lecture by Berendsen, so we will not discuss it further.

In PBC we include in the MI potential the interaction of a particle i at $\underset{\sim}{r}_i$ with all its periodic replicas and with all the other dipoles and all their replicas. That is, we replace $\phi_{ij}^{L}(\underset{\sim}{r}_{ij}^{*})$ in eq. (1.4) by the lattice sum

$$\sum_{\underset{\sim}{n}\in\Lambda_c}^{*} q_i q_j |\underset{\sim}{r}_{ij} + \underset{\sim}{n}L|^{-1} \tag{1.5}$$

for charges, the asterisk indicating that for i = j, the $\underset{\sim}{n} = \underset{\sim}{0}$ term is omitted and Λ_c being the simple cubic lattice with unit spacing and

$$\sum_{n\in\Lambda_c}^{*} \underset{\sim}{\mu}_i \cdot \left[I - \frac{3(\underset{\sim}{r}_{ij} + \underset{\sim}{n}L)(\underset{\sim}{r}_{ij} + \underset{\sim}{n}L)}{|\underset{\sim}{r}_{ij} + \underset{\sim}{n}L|^2} \right] \cdot \underset{\sim}{\mu}_j / |\underset{\sim}{r}_{ij} + \underset{\sim}{n}L|^3 \tag{1.6}$$

for dipoles. This means that in PBC the effective Hamiltonian for an ionic system is

$$H_{PBC}(\underset{\sim}{r}_1,\ldots,\underset{\sim}{r}_N) = \frac{1}{2} \sum_{i=1}^{N} \sum_{\substack{j=1 \\ i\neq j}}^{N} \phi_{ij}^{S}(\underset{\sim}{r}_{ij}^{*})$$

$$+ \frac{1}{2} \sum_{n\in\Lambda_c}^{*} \sum_{i=1}^{N} \sum_{j=1}^{N} q_i q_j |\underset{\sim}{r}_{ij} + \underset{\sim}{n}L|^{-1} \tag{1.7}$$

and for a dipolar system

$$H_{PBC}(\underset{\sim}{r}_1,\ldots,\underset{\sim}{r}_N) = \frac{1}{2} \sum_{i=1}^{N} \sum_{\substack{j=1 \\ i\neq j}}^{N} \phi_{ij}^{S}(\underset{\sim}{r}_{ij}^{*})$$

$$- \frac{1}{2} \sum_{n\,\Lambda_c}^{*} \sum_{i=1}^{N} \sum_{j=1}^{N} (\underset{\sim}{\mu}_i \cdot \nabla)(\underset{\sim}{\mu}_j \cdot \nabla) \frac{1}{|\underset{\sim}{r}|} \bigg|_{\underset{\sim}{r}=\underset{\sim}{r}_{ij}+L\underset{\sim}{n}} \tag{1.8}$$

The lattice sums in eqs. (1.7 and 1.8) have been discussed at some length in an earlier lecture at this conference. They are conditionally convergent (in the ionic case only because of the overall neutrality constraint

$$\sum_{i=1}^{N} q_i = 0 \tag{1.9}$$

unless the dipole moment of the whole configuration

$$\underset{\sim}{M} = \sum_{i=1}^{N} q_i \underset{\sim}{r}_i \qquad \text{or} \qquad \underset{\sim}{M} = \sum_{i=1}^{N} \underset{\sim}{\mu}_i \tag{1.10}$$

under discussion is zero. Thus we must define the order in which the sum is added up. De Leeuw, Perram and Smith [2] have discussed this problem at some length for the case of addition by spherical shells. They obtain, for the charged case

$$\sum_{\underset{\sim}{n}\in\Lambda_c}^{*} \sum_{i=1}^{N} \sum_{j=1}^{N} q_i q_j |\underset{\sim}{r}_{ij} + L\underset{\sim}{n}|^{-1} = \frac{1}{L} \sum_{i=1}^{N} \sum_{j=1}^{N} q_i q_j \, \psi_E\!\left(\frac{\underset{\sim}{r}_{ij}}{L}, \Lambda_c\right) + \frac{2\pi}{3L^3}\left(\sum_{i=1}^{N} q_i \underset{\sim}{r}_i\right)^2 \tag{1.11}$$

and for the dipolar case

$$\sum_{\underset{\sim}{n}\in\Lambda_c}^{*} \sum_{i=1}^{N} \sum_{j=1}^{N} \left\{ -(\underset{\sim}{\mu}_i \cdot \nabla)(\underset{\sim}{\mu}_j \cdot \nabla) \frac{1}{|\underset{\sim}{r}|} \bigg|_{\underset{\sim}{r}=\underset{\sim}{r}_{ij}+L\underset{\sim}{n}} \right\}$$
$$= \frac{1}{L^3} \sum_{i=1}^{N} \sum_{j=1}^{N} \left\{ -(\underset{\sim}{\mu}_i \cdot \nabla)(\underset{\sim}{\mu}_j \cdot \nabla) \psi_E\!\left(\frac{\underset{\sim}{r}_{ij}}{L}, \Lambda_c\right) \right\} + \frac{2\pi}{3L^3}\left(\sum_{i=1}^{N} \underset{\sim}{\mu}_i\right)^2 . \tag{1.12}$$

Here the function $\psi_E(\underset{\sim}{r}, \Lambda_c)$ is given by

$$\psi_E(\underset{\sim}{r}, \Lambda_c) = f(\underset{\sim}{r}) + \sum_{\substack{\underset{\sim}{n}\in\Lambda_c \\ \underset{\sim}{n}\neq \underset{\sim}{0}}} \frac{\operatorname{erfc}(\alpha|\underset{\sim}{r}+\underset{\sim}{n}|)}{|\underset{\sim}{r}+\underset{\sim}{n}|} + \frac{e^{-\pi^2 \underset{\sim}{n}^2/\alpha^2 + 2\pi i \underset{\sim}{n}\cdot\underset{\sim}{r}}}{\pi \underset{\sim}{n}^2} \tag{1.13}$$

with

$$f(\underset{\sim}{r}) = \begin{cases} \dfrac{\operatorname{erfc}(\alpha|\underset{\sim}{r}|)}{|\underset{\sim}{r}|}, & \underset{\sim}{r} \neq \underset{\sim}{0} \\ -2\alpha/\sqrt{\pi}, & \underset{\sim}{r} = \underset{\sim}{0} \end{cases} \tag{1.14}$$

We may evaluate the double grad operation in eq. (1.12) to obtain

$$\sum_{\underset{\sim}{n}\in\Lambda_c}^{*} \left\{ \sum_{i=1}^{N} \sum_{j=1}^{N} -(\underset{\sim}{\mu}_i \cdot \nabla)(\underset{\sim}{\mu}_j \cdot \nabla) \frac{1}{|\underset{\sim}{r}|} \Bigg|_{\underset{\sim}{r}=\underset{\sim}{r}_{ij}+L\underset{\sim}{n}} \right\}$$

$$= \frac{1}{L^3} \sum_{i=1}^{N} \sum_{j=1}^{N} \underset{\sim}{\mu}_i \cdot T(\underset{\sim}{r}_{ij}/L) \cdot \underset{\sim}{\mu}_j + \frac{2\pi}{3L^3} \left(\sum_{i=1}^{N} \underset{\sim}{\mu}_i \right)^2 , \tag{1.15}$$

where

$$T(\underset{\sim}{r}) = \sum_{\substack{\underset{\sim}{n}\in\Lambda_c \\ \underset{\sim}{n}\neq\underset{\sim}{0}}} \left\{ \left[\frac{\operatorname{erfc}(\alpha|\underset{\sim}{r}+\underset{\sim}{n}|)}{|\underset{\sim}{r}+\underset{\sim}{n}|^3} + \frac{2\alpha}{\sqrt{\pi}|\underset{\sim}{r}+\underset{\sim}{n}|^2} e^{-\alpha^2(\underset{\sim}{r}+\underset{\sim}{n})^2} \right] \right.$$

$$\times \left[I - \frac{3(\underset{\sim}{r}+\underset{\sim}{n})(\underset{\sim}{r}+\underset{\sim}{n})}{|\underset{\sim}{r}+\underset{\sim}{n}|^2} \right]$$

$$- 4\alpha^3 \frac{(\underset{\sim}{r}+\underset{\sim}{n})(\underset{\sim}{r}+\underset{\sim}{n})}{|\underset{\sim}{r}+\underset{\sim}{n}|^2} \exp(-\alpha^2(\underset{\sim}{r}+\underset{\sim}{n})^2)$$

$$\left. - 4\pi \frac{\underset{\sim}{n}\underset{\sim}{n}}{\underset{\sim}{n}^2} e^{-\pi^2 \underset{\sim}{n}^2/\alpha^2 + 2\pi i \underset{\sim}{n}\cdot\underset{\sim}{r}} \right\} + G(\underset{\sim}{r}) \tag{1.16}$$

with

$$G(\underset{\sim}{r}) = \begin{cases} \left[\dfrac{\operatorname{erfc}(\alpha|\underset{\sim}{r}|)}{|\underset{\sim}{r}|^3} + \dfrac{2\alpha}{\sqrt{\pi}|\underset{\sim}{r}|^2} e^{-\alpha^2 \underset{\sim}{r}^2} \right] \left[I - \dfrac{3\underset{\sim}{r}\underset{\sim}{r}}{|\underset{\sim}{r}|^2} \right] - \dfrac{4\alpha^3 \underset{\sim}{r}\underset{\sim}{r}}{\sqrt{\pi}\, \underset{\sim}{r}^2} e^{-\alpha^2 \underset{\sim}{r}^2} & \text{for } \underset{\sim}{r} \neq \underset{\sim}{0}, \\ -\dfrac{4\alpha^3}{3\sqrt{\pi}} I & \text{for } \underset{\sim}{r} = \underset{\sim}{0} \end{cases} \tag{1.17}$$

These expressions may be used to give a PBC Hamiltonian for the case when the lattice sums are added by spherical shells.

We may, however, add the lattice sum in a slightly different way. When we add by spherical shells we are looking at, for the ionic case

$$\lim_{R\to\infty} \sum_{\substack{\underset{\sim}{n}\in\Lambda_c;\\ |\underset{\sim}{n}|\leq R}} \sum_{i=1}^{N} \sum_{j=1}^{N} q_i q_j |\underset{\sim}{r}_{ij} + \underset{\sim}{n}L|^{-1} . \tag{1.18}$$

When we study the object in (1.18) before taking the limit, we may ask what will be the effect on the energy if the region outside the sphere of radius R be filled with a continuum dielectric of dielectric constant ε'. If the simulation cell has a net dipole moment $\underset{\sim}{M}$, then for very large R we may model this system as a sphere with uniform polarization density $\underset{\sim}{M}/L^3$ immersed in the continuum dielectric. The sphere will polarize the dielectric, thus setting up a reaction field which interacts with the polarization density. Half of this interaction energy should be associated with the simulation cells and half with the continuum. In the limit $R \to \infty$, De Leeuw, Perram and Smith [2] have showed that this energy is

$$-\frac{2(\varepsilon' - 1)}{2\varepsilon' + 1} \cdot \frac{2\pi}{3L^3} \underset{\sim}{M}^2 \tag{1.19}$$

per simulation cell. Thus the PBC Hamiltonian depends on the choice of external dielectric constant ε', and we talk in future of PBC(ε').

If we put these calculations together we obtain the PBC Hamiltonians:

(i) For ionic systems

$$H_{PBC(\varepsilon')}(\underset{\sim}{r}_1,\ldots,\underset{\sim}{r}_N) = \frac{1}{2} \sum_{\substack{i=1\\ i\neq j}}^{N} \sum_{j=1}^{N} \left\{\phi_{ij}{}^{S}(\underset{\sim}{r}_{ij}^{*}) + \frac{q_i q_j}{L}\,\psi_E\left(\frac{\underset{\sim}{r}_{ij}}{L},\, \Lambda_c\right)\right\}$$

$$+ \frac{2\pi}{3L^3}\left(1 - \frac{2(\varepsilon' - 1)}{2\varepsilon' + 1}\right)\left(\sum_{i=1}^{N} q_i \underset{\sim}{r}_i\right)^2 . \tag{1.20}$$

(ii) For dipolar systems

$$H_{PBC(\varepsilon')}(\underset{\sim}{r}_1,\ldots,\underset{\sim}{r}_N) = \frac{1}{2}\sum_{\substack{i=1\\i\neq j}}^{N}\sum_{j=1}^{N}\left\{\phi_{ij}{}^{S}(\underset{\sim}{r}_{ij}{}^{*}) + \frac{1}{L^3}\underset{\sim}{\mu}_i\cdot T\left(\frac{\underset{\sim}{r}_{ij}}{L}\right)\cdot\underset{\sim}{\mu}_j\right\}$$

$$+ \frac{2\pi}{3L^3}\left(1 - \frac{2(\varepsilon' - 1)}{2\varepsilon' + 1}\right)\left(\sum_{i=1}^{N}\underset{\sim}{\mu}_i\right)^2 . \qquad (1.21)$$

We shall now discuss methods of using these lattice sums in simulations and also the effects of the choice of ε' on the simulatio We note that the charge-charge interaction in PBC(ε') may be written

$$\frac{q_i q_j}{L}\,\psi_E(\underset{\sim}{r}_{ij}/L) - \frac{kT}{L^3}\,\nu(\varepsilon',1)\Big[z_i z_j \underset{\sim}{r}_i\cdot\underset{\sim}{r}_j$$

$$+ \frac{1}{2(N-1)}\,(z_i^2\underset{\sim}{r}_i^2 + z_j^2\underset{\sim}{r}_j^2)\Big] \qquad (1.22)$$

where $q_i = z_i Q$ and z_i is a positive or negative integer, and

$$\nu(\varepsilon'',\varepsilon') = \frac{4\pi}{3}\frac{Q^2}{kT}\left[\frac{2(\varepsilon''-1)}{2\varepsilon''+1} - \frac{2(\varepsilon'-1)}{2\varepsilon'+1}\right] . \qquad (1.23)$$

On the other hand the dipole-dipole interaction may be written

$$\frac{1}{L^3}\,\underset{\sim}{\mu}_i\cdot T(\underset{\sim}{r}_{ij}/L)\cdot\underset{\sim}{\mu}_j - \frac{kT}{N}\,\lambda(\varepsilon',1)\hat{\underset{\sim}{\mu}}_i\cdot\hat{\underset{\sim}{\mu}}_j \qquad (1.24)$$

where

$$\lambda(\varepsilon'',\varepsilon') = 3y\left[\frac{2(\varepsilon''-1)}{2\varepsilon''+1} - \frac{2(\varepsilon'-1)}{2\varepsilon'+1}\right] \qquad (1.25)$$

and

$$y = \frac{4\pi}{9}\,\frac{|\underset{\sim}{\mu}|^2 N}{kT\,L^3} \qquad (1.26)$$

These expressions are useful as they write the effective pair potential in terms of a pair potential independent of ε' plus an extra pair potential which is very weak for large L, but very long-ranged.

II. NUMERICAL EVALUATION OF LATTICE SUMS

We consider here the problem of using the function $\psi_e(\underset{\sim}{r}_{ij},\Lambda_c)$ in a simulation. It is composed of two lattice sums and for $\underset{\sim}{r}_{ij} \neq \underset{\sim}{0}$ may be written

$$\psi_E(\underset{\sim}{r}_{ij},\Lambda_c) = \sum_{\underset{\sim}{n}\in\Lambda_c} \frac{\mathrm{erfc}(\alpha|\underset{\sim}{r}_{ij}+\underset{\sim}{n}|)}{|\underset{\sim}{r}_{ij}+\underset{\sim}{n}|} + \sum_{\underset{\sim}{m}\neq\underset{\sim}{0}} \frac{e^{-\pi^2\underset{\sim}{m}^2/\alpha^2+2\pi i\underset{\sim}{m}\cdot\underset{\sim}{r}_{ij}}}{\pi\underset{\sim}{m}^2} \tag{2.1}$$

Since we must evaluate it many times in a simulation, we must consider how it may be evaluated quickly. An excellent discussion of this problem is contained in ref. [3], but since this is not widely available we discuss it here in a simple way. Suppose we wish to calculate $\psi_E(\underset{\sim}{r},\Lambda_c)$ with an error of less than δ for all $\underset{\sim}{r} \in [-\frac{1}{2},\frac{1}{2}]^3$. The erfc(x) algorithms require that an exponential and a polynomial be evaluated [4]. To minimize this work, we choose α so large that only one term in the sum of erfc's in (2.1) beed be evaluated. We may approximate the contribution of the rest by the largest omitted term. The largest that this term can be is $\mathrm{erfc}(\alpha/2)/(1/2) \simeq 4\exp(-\alpha^2/4)/\sqrt{\pi}\,\alpha$ [5]. An approximate solution to this problem is then given by

$$e^{-\alpha^2/4} \leq \sqrt{\pi}\,\delta/2 \Rightarrow \alpha > 2\left[-\log\left(\frac{\delta\sqrt{\pi}}{2}\right)\right]^{\frac{1}{2}} \tag{2.2}$$

and for $\delta = 10^{-6}$ this gives $\alpha \geq 7{\cdot}47$. As De Leeuw showed, this crude error bound is rather too conservative, and he used a value $\alpha = 4{\cdot}86$. We now turn to the other part of $\psi_E(\underset{\sim}{r},\Lambda_c)$. We must sum over a finite set of vectors $\underset{\sim}{m}$, so that if we sum over vectors $|\underset{\sim}{m}| \leq A$, the sum over the omitted vectors must satisfy the inequality

$$\sum_{|\underset{\sim}{m}|>A} \frac{e^{-\pi^2\underset{\sim}{m}^2/\alpha^2}}{\pi\underset{\sim}{m}^2} \leq \delta/2 \ . \tag{2.3}$$

We may estimate the sum on the left side of (2.2) by the integral

$$\int_{|\underset{\sim}{r}|>A} d^3\underset{\sim}{r}\, \frac{e^{-\pi^2\underset{\sim}{r}^2/\alpha^2}}{\pi\underset{\sim}{r}^2} = \frac{2\alpha}{\sqrt{\pi}}\,\mathrm{erfc}(\pi A/\alpha) \ . \tag{2.4}$$

Thus, as we decrease the allowed error δ, the value of α required to make the first term in the sum of erfc's accurate increases. This means that the number of lattice vectors required also increases and for $\delta = 10^{-6}$ eqs. (2.3,4) give, approximately

$$A \geq 9{\cdot}78 \ . \tag{2.5}$$

This value of A is in fact unrealistically high partly because of the crudeness of the error estimates and partly because the error estimates are for the worst possible error, a case which arises only rarely. De Leeuw discusses [3] error estimation for these sums for a range of configurations of 324 ionic particles and for $\alpha = 4{\cdot}86$ finds that A = 5 is sufficient. Values of A of 3, 4 or 5 are commonly used. The main usefulness of the error bounds discussed here is to show how the choice of α and the choice of the number of lattice vectors in the second sum in eq. (2.1) jointly affect the error in the resulting potentials. In his discussion of Monte Carlo simulation of dipolar systems [6], Adams notes that some results can depend crucially on the accuracy of the potential calculations: for dipolar systems the calculation of the dielectric constant seems to require more lattice vectors in this sum than might otherwise have been expected.

The function $\operatorname{erfc}(\alpha|\underset{\sim}{r}|)/|\underset{\sim}{r}|$ must still be evaluated once for each ion pair. De Leeuw [7] recommends storing a table of values of $\operatorname{erfc}(\alpha|\underset{\sim}{r}|)/|\underset{\sim}{r}|$ as a function of $\underset{\sim}{r}^2$ and using linear interpolation from it. This method of tabulation can mean that it is not necessary to evaluate $|\underset{\sim}{r}|$, which gives an added saving of time.

This procedure leaves us with the problem of evaluating

$$\sum_{\underset{\sim}{m}\neq\underset{\sim}{0}} \frac{e^{-\pi^2\underset{\sim}{m}^2/\alpha^2}}{\pi\underset{\sim}{m}^2} e^{2\pi i\underset{\sim}{m}\cdot\underset{\sim}{r}_{ij}} \; . \tag{2.6}$$

Singer [8] has suggested that it is much more efficient to calculate the contribution for all i,j pairs of these terms, namely

$$\sum_{i=1}^{N}\sum_{j=1}^{N}\sum_{\underset{\sim}{m}\neq\underset{\sim}{0}} \frac{e^{-\pi^2\underset{\sim}{m}^2/\alpha^2}}{\pi\underset{\sim}{m}^2} e^{2\pi i\underset{\sim}{m}\cdot\underset{\sim}{r}_{ij}}$$

$$= \sum_{\underset{\sim}{m}\neq\underset{\sim}{0}} \frac{e^{-\pi^2\underset{\sim}{m}^2/\alpha^2}}{\pi\underset{\sim}{m}^2} \left|\sum_{i=1}^{N} e^{2\pi i\underset{\sim}{m}\cdot\underset{\sim}{r}_{i}}\right|^2 \; . \tag{2.7}$$

This reduces the problem from a sum over N(N-1)/2 pairs to a sum of 2N terms $\left(\sum_{i=1}^{N}\cos(2\pi\underset{\sim}{m}\cdot\underset{\sim}{r}_i) + \sum_{i=1}^{N}\sin(2\pi\underset{\sim}{m}\cdot\underset{\sim}{r}_i)\right)$ and so drastically reduces the required computing time. Perram [9] (and no doubt several others) recommend storing a table of values of $\exp(-\pi^2\underset{\sim}{m}^2/\alpha^2)/\pi\underset{\sim}{m}^2$. It may also be found that since the sine and cosine algorithms for a given machine are designed to cope with any real argument, it may be worth writing a separate algorithm which has arguments only in the range $[-\pi/2,\pi/2]$, since this is the only range required. Such an algorithm may well be faster for the values of the argument required than a built-in machine algorithm.

III. EFFECTS OF CHOICE OF EXTERNAL DIELECTRIC CONSTANT

In using the boundary conditions PBC(ε') we include in the Hamiltonian a small long-ranged potential which depends on the value of ε'. In a simulation we want to calculate static and dynamic effects and in this lecture I will confine myself to discussing the effects on the static properties of the system. We shall discuss the internal energy, the mean square fluctuations of the dipole moment of the system, the static dielectric constant and the correlation functions of the system. First we discuss a dipolar system. We present results derived in refs. [2 and 10]. For a system of molecules with a dipole moment per particle and linear symmetry, the two particle correlation function may be written in the standard forms

$$g(1,2) = \sum_{\ell}\sum_{\ell'}\sum_{m} g_{\ell\ell' m}(|\underset{\sim}{r}_{12}|)Y_{\ell}^{m}(1)Y_{\ell'}^{-m}(2), \tag{3.1}$$

and

$$g(1,2) = g_R(r_{12}) + h_D(r_{12})D(1,2) + h_\Delta(r_{12})\Delta(1,2) + \cdots . \tag{3.2}$$

The Kirkwood g-factor g_K, which is the mean square polarization per particle in the system is defined as

$$g_K = \left[< (\sum_{i=1}^{N} \underset{\sim}{\mu}_i)^2 > - < \sum_{i=1}^{N} \underset{\sim}{\mu}_i >^2\right]/N|\underset{\sim}{\mu}|^2 . \tag{3.3}$$

For a large sphere containing N particles at density ρ, with $y = 4\pi\rho\mu^2/9\ kT$, and with the environment external to the sphere being a dielectric continuum of dielectric constant ε', g_K is related to the dielectric constant ε of the sphere by

$$\frac{(\varepsilon-1)(2\varepsilon'+1)}{3(\varepsilon+2\varepsilon')} = yg_K(\varepsilon') \tag{3.4}$$

where we assume that g_K deoends on '. If it did not so depend, then the dielectric constant ε would have to depend on ε', and this is not an acceptable assumption. The internal energy may be calculated in the standard way [10] from $h_D(r_{12})$. The Kirkwood g-factor depends on $h_\Delta(r)$ via

$$g_K(\varepsilon') = 1 + \frac{\rho}{3}\int d^3\underset{\sim}{r}\ h_{\Delta\varepsilon'}(|\underset{\sim}{r}|) \tag{3.5}$$

where h_Δ has the added subscript ε' to show where the dependence of $g_K(\varepsilon')$ on ε' comes from. The interparticle potential may be written in the form

$$\phi_{\varepsilon'}(1,2) = \phi^S(1,2) + \frac{1}{L^3}\,\underset{\sim}{\mu}_1\cdot T(\underset{\sim}{r}_{12}/L)\cdot\underset{\sim}{\mu}_2 - \frac{kT}{N}\,3y\left(\frac{2(\varepsilon'-1)}{2\varepsilon'+1} - 1\right)\hat{\underset{\sim}{\mu}}_1\cdot\hat{\underset{\sim}{\mu}}_2 \qquad (3.6)$$

so that

$$\phi_{\varepsilon''}(1,2) = \phi^S(1,2) + \frac{1}{L^3}\,\underset{\sim}{\mu}_1\cdot T(\underset{\sim}{r}_{12}/L)\cdot\underset{\sim}{\mu}_2 - \frac{kT}{N}\,\lambda(\varepsilon'',\varepsilon')\hat{\underset{\sim}{\mu}}_1\cdot\hat{\underset{\sim}{\mu}}_2 \quad . \qquad (3.7)$$

Notice that if $\varepsilon' = 1$, then the term in $\hat{\underset{\sim}{\mu}}_1\cdot\hat{\underset{\sim}{\mu}}_2$ is $3y\,kT\,\hat{\underset{\sim}{\mu}}_1\cdot\hat{\underset{\sim}{\mu}}_2/N$ and

$$\frac{\varepsilon-1}{\varepsilon+2} = yg_K(1) \quad , \qquad (3.8)$$

the usual Clausius-Mosotti formula. If $\varepsilon' \to \infty$, the term in $\hat{\underset{\sim}{\mu}}_1\cdot\hat{\underset{\sim}{\mu}}_2$ vanishes from the Hamiltonian and

$$\varepsilon = 1 + 3y\,g_K(\infty) \quad . \qquad (3.9)$$

This limit may arise if we consider the sphere of material to be surrounded by a metal. If $\varepsilon' = \varepsilon$, then we obtain the Kirkwood formula

$$\frac{(\varepsilon-1)(2\varepsilon+1)}{9\varepsilon} = y\,g_K(\varepsilon) \quad . \qquad (3.10)$$

The terms in $\hat{\underset{\sim}{\mu}}_1\cdot\hat{\underset{\sim}{\mu}}_2$ in the PBC(ε') potential in eqs. (3.6 and 7) are very small but long-ranged. De Leeuw, Perram and Smith [10] have treated this potential as a perturbation and using graphical perturbation techniques [11] have summed the infinite class of graphs which contribute to g(1,2) to order 1/N, neglecting terms which are $O(1/N^2)$. They are thus able to see the effects of changing the external dielectric constant from ε' to ε''. A summary of their results is

(i) $g_{R\varepsilon'}(r_{12})$ is shifted by a term proportional to $h_\Delta(r_{12})/N$.
(ii) $h_{D\varepsilon'}(r_{12})$ is shifted by a term which is $O(1/N^2)$.
(iii) $h_{\Delta\varepsilon'}(r_{12})$ is shifted by a term proportional to $g_{R\varepsilon'}(r_{12})/N$ which tends to 1/N for large r_{12}, and does not decay with distance. Thus at large r_{12}, $h_{\Delta\varepsilon'}(r_{12})$ is negative and constant if $\varepsilon' < \varepsilon$ and positive and constant if $\varepsilon' > \varepsilon$.
(iv) The internal energy is shifted by an amount which is $O(1/N^2)$.
(v) In the limit $N \to \infty$,

$$g(\varepsilon'') = g(\varepsilon')/(1-\lambda(\varepsilon'',\varepsilon')g(\varepsilon')/3) \qquad (3.11)$$

(vi) Eqs. (3.11) and (3.4) may be used to prove that ε is independent of ε'.

These points are discussed further by Perram [12] and Adams [13] in their lectures.

For an ionic system, similar considerations apply. The effective pair potential may be written

$$\psi_{ij}(\underset{\sim}{r}_i,\underset{\sim}{r}_j;\varepsilon') = \phi_{ij}{}^{S}(\underset{\sim}{r}_{ij}) + \frac{q_i q_j}{L}\,\psi_E(\underset{\sim}{r}_{ij}/L,\Lambda_c)$$

$$- \frac{kT}{L^3}\left(\frac{2(\varepsilon'-1)}{2\varepsilon'+1} - 1\right)\cdot\frac{4\pi Q^2}{3kT}\left[z_i z_j\,\underset{\sim}{r}_i\cdot\underset{\sim}{r}_j + \frac{1}{2(N-1)}\right.$$

$$\left.\times\left[z_i{}^2\underset{\sim}{r}_i{}^2 + z_j{}^2\underset{\sim}{r}_j{}^2\right]\right] \tag{3.12}$$

so that

$$\psi_{ij}(\underset{\sim}{r}_i,\underset{\sim}{r}_j;\varepsilon'') = \psi_{ij}(\underset{\sim}{r}_i,\underset{\sim}{r}_j;\varepsilon')$$

$$- \frac{kT\,\nu(\varepsilon'',\varepsilon')}{L^3}\left[z_i z_j\,\underset{\sim}{r}_i\cdot\underset{\sim}{r}_j + \frac{1}{2(N-1)}\left[z_i{}^2\underset{\sim}{r}_i{}^2 + z_j{}^2\underset{\sim}{r}_j{}^2\right]\right] \tag{3.13}$$

We may consider the effect of changing the external dielectric constant from ε' to ε on the internal energy and the mean square polarization of the sample defined by

$$G(\varepsilon') = \left\{<\left(\sum_{i=1}^{N} z_i\underset{\sim}{r}_i\right)^2> - <\sum_{i=1}^{N} z_i\underset{\sim}{r}_i>^2\right\}/N\,. \tag{3.14}$$

Smith, Hoskins and Wright [14] have developed a perturbation theory for the effects of change in ε'. They find that the internal energy is unchanged and that in the limit $N \to \infty$,

$$G(\varepsilon'') = G(\varepsilon')\;/\;(1-\bar{\rho}G(\varepsilon')\nu(\varepsilon'',\varepsilon')/3) \tag{3.15}$$

where $\bar{\rho}$ is the number density of particles in the system.

IV. SOME SIMULATION RESULTS

We report results of Monte Carlo simulations studies of 216 hard sphere particles with embedded point charges $\pm e$, at reduced density $\eta = 0{\cdot}35$ and plasma parameter

$$\Gamma = \frac{e^2}{kT}\,(3\eta^{1/3}/\sigma) \tag{4.1}$$

(where σ is the diameter of the hard spheres) having values 2, 5 and 10 [14]. The simulations were carried out with $\varepsilon' = 1$ and $\varepsilon' \to \infty$. The system was equilibrated from a lattice of CsCl structure for $6 \times 10^5 - 1{\cdot}2 \times 10^6$ Monte Carlo moves, and averages taken over a subsequent $6 \times 10^5 - 1{\cdot}2 \times 10^6$ moves. The correlation functions were only very slightly affected by the change in ε' and the internal energy not at all, these objects looking exactly like those found by other workers when using $\varepsilon' \to \infty$ [15,16]. In fig. 1 we present $G(\infty)$. The error bars were calculated by dividing the run up into blocks of 25×10^3 configurations and calculating the root mean square deviation of the subaverages over the blocks. From the values of $G(\infty)$ it is possible to predict $G(1)$ using eq. (3.15) and the predicted and measured values of $G(1)$ are presented in the table. The formula of eq. (3.15) is seen to work fairly well.

Valleau and Whittington [17] have objected to the use of PBC because they claim that the periodicity inherent in the PBC system will distort the structure. They point to known small differences between MI and PBC(∞) simulations [18] as evidence of this distortion. We have also carried out some preliminary simulations of 64 hard sphere particles with embedded point charges at $\Gamma = 2$, 10, 15 and 20 with $\eta = 0{\cdot}35$, using MI, PBC(1) and PBC(∞) boundary conditions [19]. We found no discernible difference between the structures found in MI and PBC(1) but they were both slightly different from PBC(∞). Thus the differences between PBC(∞) and MI simulations of charged systems are not due to the periodicity but rather to the long-ranged term

$$-\frac{kT}{L^3}\,\nu(\infty,1)\left\{z_i z_j\,\underset{\sim}{r}_i\cdot\underset{\sim}{r}_j + \frac{1}{2(N-1)}\left[z_i^2\underset{\sim}{r}_i^2 + z_j^2\underset{\sim}{r}_j^2\right]\right\} \tag{4.2}$$

in the effective charge-charge potential of eq. (3.13). The perturbation theory of ref. [14] goes some way towards understanding the role of this term. De Leeuw [20] has pointed out that in an infinite system, $G(\infty) \to \infty$ so that

$$G(\varepsilon') = \frac{-3}{\bar{\rho}\nu(\varepsilon',\infty)} = \frac{3(2\varepsilon'+1)}{4\pi\bar{\rho}Q^2/kT}\ . \tag{4.3}$$

Some of the consequences of this relation and its connection with the Stillinger-Lovett second moment condition are discussed in his lecture. Following De Leeuw's suggestion we may insert this average value of $\underset{\sim}{M}^2/N$ into the Hamiltonian and we find if

$$\underset{\sim}{M}^2(\varepsilon') = \frac{3(2\varepsilon'+1)N}{4\pi\bar{\rho}Q^2/kT}\ , \tag{4.4}$$

then

$$-\frac{2\pi}{3L^3}\left(\frac{2(\varepsilon'-1)}{2\varepsilon'+1} - 1\right)\cdot\underset{\sim}{M}^2(\varepsilon') = \frac{3kT}{2} \tag{4.5}$$

independent of ε'. This suggests that in some sense the extra terms in the Hamiltonian for charges proportional to $\nu(\varepsilon',\infty)$ correspond to a single vibrational mode of the whole system which is independent of other motions in the system and contains an energy $3kT/2$.

Γ	ε'	$\frac{G(\varepsilon')}{\sigma^2}$ (Measured)	$\frac{G(1)}{\sigma^2}$ (Predicted)
2	1	0•65 ± 0•03	0•82 ± 0•10
2	∞	3•1 ± 1•1	
5	1	0•28 ± 0•06	0•39 ± 0•07
5	∞	2•9 ± 1•4	
10	1	0•15 ± 0•03	0•18 ± 0•02
10	∞	2•0 ± 1•0	

Table of values of $G(\varepsilon')$ for Monte Carlo simulations of 216 hard sphere particles with embedded point charges at reduced total density $\eta = 0{\cdot}35$.

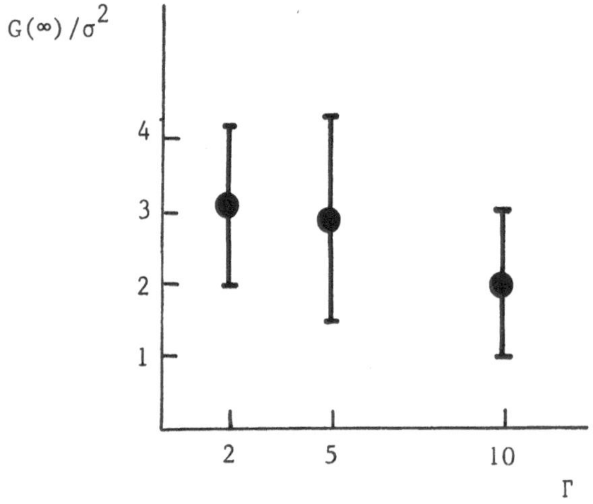

Fig. 1. Values of $G(\infty)/\sigma^2$ measured in PBC(∞) simulation.

REFERENCES

1. Barker, J.A. and Watts, R.O., Mol. Phys. 26, 789 (1973).
2. De Leeuw, S.W., Perram, J.W. and Smith, E.R., Proc. Roy. Soc. (Lond.) A373, 27 (1980).
3. De Leeuw, S.W., "Computer Simulation of the Alkaline Earth Halides", Ph.D. thesis, University of Amsterdam, 1976.
4. Abramowitz, M. and Stegun, I.A., "Handbook of Mathematical Functions", Dover, New York (1965), p. 298.
5. Gradshteyn, I.S. and Ryzhik, I.M., "Table of Integrals, Series and Products", Academic Press, New York (1965), p. 931.
6. Adams, D.J., Mol. Phys. In press.

7. De Leeuw, S.W., Private communication.
8. Singer, K., Private communication to S.W. de Leeuw and others: See ref. 22 of [3].
9. Perram, J.W., Private communication.
10. De Leeuw, S.W., Perram, J.W. and Smith, E.R., Proc. Roy. Soc. (Lond.) A373, 57 (1980).
11. Smith, E.R. and Perram, J.W., Match 4, 3 (1978).
12. Perram, J.W., Lecture at this wirkshop (ch. 8).
13. Adams, D.J., Lecture at this workshop (Ch. 9).
14. Smith, E.R., Hoskins, C.S. and Wright, C.C., Mol. Phys. In press.
15. Larsen, B., Chem. Phys. Letts. 27, 47 (1974).
16. Adams, D.J., Chem. Phys. Letts. 62, 329 (1979).
17. Valleau, J.P. and Whittington, S., "Statistical Mechanics. Part A. Equilibrium Methods", ed. B.J. Berne (Plenum, New York, 1977).
18. Hoskins, C.S. and Smith, E.R., Research Report #19, Mathematics Department, University of Newcastle (1978).
19. Hoskins, C.S. and Smith, E.R., Mol. Phys. 41, 243 (1980).
20. De Leeuw, S.W., Private communication.

EVALUATION OF LATTICE SUMS IN DISORDERED IONIC SYSTEMS

Edgar R. Smith[a] and John W. Perram[b]

[a]Department of Mathematics
University of Melbourne
Parkville, Vic. 3052
Australia

and

[b]Matematisk Institut
Odense Universitet
Campusvej 55
5230 Odense M
Denmark

ABSTRACT

The potential energy of a crystal of point ions in an external field or the field set up by an impurity is written down and from it a set of equations for the distortion of the lattice caused by the field are derived. Since the distortions may often be small, it is clear that solutions of the linearized equations will be very useful. The linearized equations may be studied by Fourier transform techniques. These techniques require that lattice sums for ion-ion and ion-defect potentials be analysed with some care. The relevant sums are analysed in this paper, in a form which allows their use in some lattice distortion problems.

I. INTRODUCTION

We consider a crystal made up of point ions. The energy of a large piece of crystal contains an electrostatic term which depends on the shape of the piece of crystal, the lattice vectors, the distribution of the ions within the unit cell and the relevant charge interaction. Accordingly we consider an infinite lattice L. The lattice vectors $\underset{\sim}{n} \in L$ have the form

$$\underset{\sim}{n} = n_1\underset{\sim}{a}_1 + n_2\underset{\sim}{a}_2 + n_3\underset{\sim}{a}_3 \,, \tag{1.1}$$

with n_i being any integer and the lattice vectors define the centres of the unit cells of the crystal. The reciprocal lattice R is composed of lattice vectors

$$\underset{\sim}{m} = m_1\underset{\sim}{A}_1 + m_2\underset{\sim}{A}_2 + m_3\underset{\sim}{A}_3 \tag{1.2}$$

where the m_j are any integers,

$$\underset{\sim}{A}_1 = \frac{\underset{\sim}{a}_2 \times \underset{\sim}{a}_3}{|L|} \,, \quad \underset{\sim}{A}_2 = \frac{\underset{\sim}{a}_3 \times \underset{\sim}{a}_1}{|L|} \,, \quad \underset{\sim}{A}_3 = \frac{\underset{\sim}{a}_1 \times \underset{\sim}{a}_2}{|L|} \tag{1.3}$$

and

$$|L| = |\underset{\sim}{a}_1 \cdot \underset{\sim}{a}_2 \times \underset{\sim}{a}_3| \tag{1.4}$$

is the volume of a unit cell of L. For future reference we define a reciprocal lattice space region V_x by

$$\underset{\sim}{k} \in V_x \Rightarrow \underset{\sim}{k} = \sum_{i=1}^{3} k_i\underset{\sim}{A}_i \quad \text{with} \quad -x \leq k_i \leq x, \quad i = 1,2,3 \,. \tag{1.5}$$

To introduce the idea of shape dependence we consider the region P_1 of real space which is all points inside or on the surface given by the equation

$$P(\underset{\sim}{r}) = 0 \,. \tag{1.6}$$

A large P-shaped region is then P_N which is all points inside or on the surface given by

$$P(\underset{\sim}{r}/N) = 0 \,. \tag{1.7}$$

The limit of a large P-shaped region is then obtained by considering P_N in the limit $N \to \infty$. The finite, large subset $P_N(L)$ of the lattice L is then all lattice vectors $\underset{\sim}{n} \in L$ for which

$$\underset{\sim}{n} \in P_N \,.$$

The two examples of shapes to be considered here are a sphere, for which

$$P(\underset{\sim}{r}) = \underset{\sim}{r}^2 = 1 \tag{1.8}$$

and a plane slab normal to the z-direction for which

$$P(\underset{\sim}{r}) = [\underset{\sim}{r}\cdot(0,0,1)]^2 = 1 \quad . \tag{1.9}$$

For the spherical case the shape P_N is thus a sphere of radius N. For the planar case, the shape P_N is a slab of width 2N.

The unit cell of an ionic lattice contains H point ions at position $\underset{\sim}{r}_1,\ldots,\underset{\sim}{r}_H$, bearing charges $q_1,\ldots,q_H$. The unit cell, with centre at $\underset{\sim}{n}$, has a particle of type i at $\underset{\sim}{r}_i + \underset{\sim}{n}$. The total charge in a unit cell is zero:

$$\sum_{i=1}^{H} q_i = 0 \quad . \tag{1.10}$$

An object of great interest is the dipole moment per unit cell

$$\underset{\sim}{M} = \sum_{i=1}^{H} q_i \underset{\sim}{r}_i \quad . \tag{1.11}$$

In addition to the normal Coulomb interaction between charges, there will also be a set of short-ranged interactions which we shall write $\phi_{ij}(\underset{\sim}{r}_{ij})$ for ions of type i and j. The total interaction between ions of type i and j is then

$$\phi_{ij}^{T}(\underset{\sim}{r}_{ij}) = \phi_{ij}(\underset{\sim}{r}_{ij}) + q_i q_j/|\underset{\sim}{r}_{ij}| \quad . \tag{1.12}$$

We may note that polarizability effects may be accounted for by considering each ion to be a coupled ion pair.

The total energy of a piece $P_N(L)$ of the crystal may be written

$$W_N(L,P) = \frac{1}{2} \sum_{\underset{\sim}{n}\in P_N(L)} \sum_{\underset{\sim}{n}'\in P_N(L)} \sum_{i=1}^{H} \sum_{j=1}^{H} \phi_{ij}^{T}(\underset{\sim}{r}_i+n-\underset{\sim}{r}_j-n') \quad . \tag{1.13}$$

In eq. (1.13), the asterisk means that terms corresponding to $i = j$, $\underset{\sim}{n} = \underset{\sim}{n}'$ are to be omitted, and the factor ½ accounts for the double counting implied in the sums over i and j. It may be expected that $E_N(L)$ should have the asymptotic expansion

$$E_N(L) = AN^3 + O(N^2) \tag{1.14}$$

and we shall see that this is so. However, if $\underset{\sim}{M} \neq \underset{\sim}{0}$, it is possible to alter A in eq. (1.14) merely by changing the surface structure of the piece P_N of crystal.

To see why this is so, consider the electrostatic energy of one unit cell at the centre of P_N. It has the form

$$W_N(L,P) = \sum_{\underset{\sim}{n} \in P_N(L)} \sum_{i=1}^{H} \psi_i(\underset{\sim}{r}_i + \underset{\sim}{n} + \underset{\sim}{s}_i(\underset{\sim}{n}))$$

$$+ \frac{1}{2} \sum_{\underset{\sim}{n} \in P_N(L)} \sum^{*}_{\underset{\sim}{n}' \in P_N(L)} \sum_{i=1}^{H} \sum_{j=1}^{H} \phi_{ij}{}^{T}(\underset{\sim}{r}_i + \underset{\sim}{n} + \underset{\sim}{s}_i(\underset{\sim}{n})$$

$$- \underset{\sim}{r}_j - \underset{\sim}{n}' - \underset{\sim}{s}_j(\underset{\sim}{n}')) \ . \tag{1.16}$$

The equilibrium distortion of the crystal is given by the solution $\underset{\sim}{s}_k(\underset{\sim}{m})$ of the equation

$$\nabla_{\underset{\sim}{s}_k(\underset{\sim}{m})} W_N(L,P) = \underset{\sim}{0} \ , \quad k = 1,\ldots,H; \quad \underset{\sim}{m} \in P_N(L) \ . \tag{1.17}$$

One feature of these equations which can be of use is to note that if $\psi_k(\underset{\sim}{r}) = 0$ there exists k (the crystal is undistorted), then $\underset{\sim}{s}_k(\underset{\sim}{m}) = \underset{\sim}{0}$ so that

$$\sum^{*}_{\underset{\sim}{n} \ P_N(L)} \sum_{j=1}^{H} \nabla\phi_{kj}{}^{T}(\underset{\sim}{r}_k + \underset{\sim}{m} - \underset{\sim}{r}_j - \underset{\sim}{n}) = 0, \quad k = 1,\ldots,H; \ \underset{\sim}{m} \in P_N(L) \ . \tag{1.18}$$

We may write eq. (1.17) explicitly in the form

$$\nabla\psi_k(\underset{\sim}{r}_k + \underset{\sim}{m} + \underset{\sim}{s}_k(\underset{\sim}{m})) + \sum^{*}_{\underset{\sim}{n} \in P_N(L)} \sum_{j=1}^{H} \nabla\phi_{kj}{}^{T}(\underset{\sim}{r}_{kj} + \underset{\sim}{m} - \underset{\sim}{n} + \underset{\sim}{s}_k(\underset{\sim}{m}) - \underset{\sim}{s}_j(\underset{\sim}{n})) = 0. \tag{1.19}$$

These equations are non-linear in the $\underset{\sim}{s}_k(\underset{\sim}{m})$. They require proper evaluation of, among other things

$$\theta(\underset{\sim}{r};L,P,N) = \sum_{\underset{\sim}{n} \in P_N(L)} |\underset{\sim}{r} + \underset{\sim}{n}|^{-1} \tag{1.20}$$

and its gradient. If we presume that the distortions $\underset{\sim}{s}_k(\underset{\sim}{m})$ are all small, then eq. (1.19) may be linearized to give

$$\nabla\psi_k(\underset{\sim}{r}_k + \underset{\sim}{m}) + (\underset{\sim}{s}_k(\underset{\sim}{m}) \cdot \nabla)\nabla\psi_k(\underset{\sim}{r}_k + \underset{\sim}{m})$$

$$+ \sum^{*}_{\underset{\sim}{n} \in P_N(L)} \sum_{j=1}^{H} (\underset{\sim}{s}_k(\underset{\sim}{m}) - \underset{\sim}{s}_j(\underset{\sim}{n}) \ \cdot\nabla)\nabla\phi_{kj}{}^{T}(\underset{\sim}{r}_{kj} + \underset{\sim}{m} - \underset{\sim}{n}) = \underset{\sim}{0} \ . \tag{1.21}$$

These equations may be Fourier transformed. We define the Fourier transform of a function by

$$E_N(L,P) = \frac{1}{2} \sum_{\underset{\sim}{n}}^{*} \sum_{P_N(L)} \sum_{i=1}^{H} \sum_{j=1}^{H} q_i q_j |\underset{\sim}{r}_{ij} + \underset{\sim}{n}|^{-1} \quad . \tag{1.15}$$

Eq. (1.15) contains a sum of electrostatic interactions of the ions in the cell with centre at $\underset{\sim}{0}$ with those with centre at $\underset{\sim}{n}$. If the unit cell has no dipole moment, then the sum on $\underset{\sim}{n}$ contains terms which are at the largest, $O(|\underset{\sim}{n}|^{-5})$ for large $|\underset{\sim}{n}|$, so that the sum on $\underset{\sim}{n}$, in the limit $N \to \infty$ is absolutely convergent. If $M \neq 0$, there are terms which are $O(|\underset{\sim}{n}|^{-3})$ for large $|\underset{\sim}{n}|$, so that the sum on $\underset{\sim}{n}$ can only be conditionally convergent for $N \to \infty$ [1]. Thus, the order of summation, given by the shape of P_N will determine in part the answer obtained. It is pertinent to consider some model crystals.

(i) The CsCl structure. The lattice vectors are

$$\underset{\sim}{a}_1 = (1,0,0) = \underset{\sim}{A}_1,\ \underset{\sim}{a}_2 = (0,1,0) = \underset{\sim}{A}_2,\ \underset{\sim}{a}_3 = (0,0,1) = \underset{\sim}{A}_3 \ .$$

Two models may be constructed:

(a) $+\frac{q}{8}$ at $(\pm\frac{1}{2},\pm\frac{1}{2},\pm\frac{1}{2})$, $-q$ at $(0,0,0) \Rightarrow \underset{\sim}{M} = \underset{\sim}{0}$.

(b) $+q$ at $(\frac{1}{4},\frac{1}{4},\frac{1}{4})$, $-q$ at $-(\frac{1}{4},\frac{1}{4},\frac{1}{4}) \Rightarrow \underset{\sim}{M} = \frac{1}{2}q(1,1,1)$.

These two structures differ only in their surfaces, in that structure (a) must have a surface composed of fractional ions.

(ii) The fluorite structure. The lattice vectors are (1,0,0), (0,1,0), (0,0,1). Again, two models may be constructed.

(a) $\frac{q}{4}$ at $(\pm\frac{1}{2},\pm\frac{1}{2},\pm\frac{1}{2})$, $\frac{q}{2}$ at $(\pm\frac{1}{2},0,0)$, $(0,\pm\frac{1}{2},0)$, $(0,0,\pm\frac{1}{2})$

$-q$ at $(\pm\frac{1}{4},\pm\frac{1}{4},\pm\frac{1}{4}) \Rightarrow \underset{\sim}{M} = 0$.

(b) $2q$ at $(-\frac{3}{8},-\frac{3}{8},-\frac{3}{8})$, $(-\frac{3}{8},\frac{1}{8},\frac{1}{8})$, $(\frac{1}{8},-\frac{3}{8},\frac{1}{8})$, $(\frac{1}{8},\frac{1}{8},-\frac{3}{8})$.

$-q$ at $(\pm\frac{1}{4},\pm\frac{1}{4},\pm\frac{1}{4}) + \frac{1}{8}(1,1,1) \Rightarrow \underset{\sim}{M} = -2q(1,1,1)$.

The two structures differ only in their surface structures.

We may now consider the effect on $W_N(L,P)$ of including an external field or set of defects which interact with a k-ion with potential energy $\psi_k(\underset{\sim}{r})$, and which have an interaction W_D with each other in the case of a set of defects. Under the influence of these interactions, a particle at $\underset{\sim}{r}_k + \underset{\sim}{n}$ will move to a new equilibrium position $\underset{\sim}{r}_k + \underset{\sim}{n} + \underset{\sim}{s}_k(\underset{\sim}{n})$. The net energy of the distorted crystal is then

$$F(\mathbf{k}) = \sum_{\mathbf{n}\in P_N(L)} f(\mathbf{n})\, e^{2i\mathbf{k}\cdot\mathbf{n}} \tag{1.22}$$

which has inverse

$$\frac{|L|}{\pi^3}\int_{V_{\pi/2}} d^3\mathbf{k}\; F(\mathbf{k})\, e^{-2i\mathbf{k}\cdot\mathbf{n}} = \begin{cases} f(\mathbf{n}) \;, & \mathbf{n} \quad P_N(L) \\ 0 \;, & \mathbf{n} \notin P_N(L) \;. \end{cases} \tag{1.23}$$

This Fourier transform has the exact convolution theorems

$$\sum_{\mathbf{n}\in P_N(L)} f(\mathbf{n})g(\mathbf{n})e^{2i\mathbf{k}\cdot\mathbf{n}} = \frac{|L|}{\pi^3}\int_{V_{\pi/2}} F(\boldsymbol{\ell})G(\mathbf{k}-\boldsymbol{\ell})d^3\boldsymbol{\ell} \tag{1.24}$$

and

$$\frac{|L|}{\pi^3}\int_{V_{\pi/2}} F(\mathbf{k})G(\mathbf{k})e^{-2i\mathbf{k}\cdot\mathbf{m}}\, d^3\mathbf{k} = \sum_{\mathbf{n}\in P_N(L)} f(\mathbf{n})g_N(\mathbf{m}-\mathbf{n}) \;. \tag{1.25}$$

In eq. (1.25)

$$g_N(\mathbf{n}) = \begin{cases} g(\mathbf{n}) \;, & \mathbf{n} \quad P_N(L) \\ 0 \;, & \mathbf{n} \notin P_N(L) \;. \end{cases} \tag{1.26}$$

In using eq. (1.25) for very large N, it is necessary to be fairly careful if $f(\mathbf{n})$ and $g(\mathbf{n})$ decay slowly with $|\mathbf{n}|$. However, if we make the definitions

$$\mathbf{Z}_k(\mathbf{k}) = \sum_{\mathbf{m}\in P_N(L)} \mathbf{s}_k(\mathbf{m})\, e^{2i\mathbf{k}\cdot\mathbf{m}} \;, \tag{1.27a}$$

$$\boldsymbol{\Psi}_k(\mathbf{k}) = \sum_{\mathbf{m}\in P_N(L)} \nabla\psi_k(\mathbf{r}_k+\mathbf{m})\, e^{2i\mathbf{k}\cdot\mathbf{m}} \;, \tag{1.27b}$$

$$D_k(\mathbf{k}) = \sum_{\mathbf{m}\in P_N(L)} \nabla\nabla\psi_k(\mathbf{r}_k+\mathbf{m})\, e^{2i\mathbf{k}\cdot\mathbf{m}} \;, \tag{1.27c}$$

$$\Phi_{kj}(\mathbf{k}) = \sum_{\mathbf{m}\in P_N(L)} \nabla\nabla\phi_{kj}(\mathbf{r}_k-\mathbf{r}_j+\mathbf{m})\, e^{2i\mathbf{k}\cdot\mathbf{m}} \tag{1.27d}$$

and

$$T_{kj}(\mathbf{k}) = \sum_{\mathbf{m}\in P_N(L)} q_k q_j[\nabla\nabla|\mathbf{r}_{kj}+\mathbf{m}|^{-1}]\, e^{2i\mathbf{k}\cdot\mathbf{m}} \tag{1.27e}$$

then the Fourier transform of the linearized distortion equation is, for large N,

$$\Psi_k(k) + \frac{|L|}{\pi^3} \int_{V_{\pi/2}} D_k(k-\ell) \cdot Z_k(\ell) d^3\ell$$

$$+ \sum_{j=1}^{H} [\Phi_{kj}(0) - T_{kj}(0)] \cdot Z_k(k)$$

$$+ \sum_{j=1}^{H} [\Phi_{kj}(k) - T_{kj}(k)] \cdot Z_j(k) = 0 . \qquad (1.28)$$

It should be noted that this Fourier transformed equation may not be solved immediately unless $D_k(k) = 0$. Further, $T_{kj}(0)$ is not defined properly in the limit as $N \to \infty$, and some care must be taken with it.

For the rest of this lecture we consider

$$\theta(r,k;P,L,N) = \sum_{n \in P_N(L)} |r+n|^{-1} e^{2ik \cdot n} . \qquad (1.29)$$

II. THE COULOMB LATTICE SUM

We attempt to obtain an asymptotic expansion of $\theta(r,k;P,L,N)$ in powers of N. First note that [2]

$$|r+n|^{-1} = \frac{1}{\sqrt{\pi}} \int_0^\infty dt\, t^{-\frac{1}{2}} e^{-t(r+n)^2} . \qquad (2.1)$$

The integral in eq. (2.1) may be split into an integral on (y^2,∞) plus an integral on $(0,y^2)$. Thus

$$|r+n|^{-1} = \frac{1}{\sqrt{\pi}} \int_{y^2}^\infty dt\, t^{-\frac{1}{2}} e^{-t(r+n)^2} + \frac{1}{\sqrt{\pi}} \int_0^{y^2} dt\, t^{-\frac{1}{2}} e^{-t(r+n)^2} . \qquad (2.2)$$

In the first integral, we use the substitution $t = s^2$, so that $t^{-\frac{1}{2}}dt = 2ds$, and then

$$|r+n|^{-1} = \frac{\operatorname{erfc}(y|r+n|)}{|r+n|} + \frac{1}{\sqrt{\pi}} \int_0^{y^2} dt\, t^{-\frac{1}{2}} e^{-t(r+n)^2} . \qquad (2.3)$$

Thus

$$\theta(r,k;P,L,N) = \theta_1(r,k;P,L,N) + \theta_2(r,k;P,L,N) \qquad (2.4)$$

where

$$\theta_1(\underset{\sim}{r},\underset{\sim}{k};P,L,N) = \sum_{\underset{\sim}{n}\in P_N(L)} \frac{\operatorname{erfc}(y|\underset{\sim}{r}+\underset{\sim}{n}|)}{|\underset{\sim}{r}+\underset{\sim}{n}|} e^{2i\underset{\sim}{k}\cdot\underset{\sim}{n}} \tag{2.5}$$

and

$$\theta_2(\underset{\sim}{r},\underset{\sim}{k};P,L,N) = \frac{1}{\sqrt{\pi}} \int_0^{y^2} dt\, t^{-\frac{1}{2}} \sum_{\underset{\sim}{n}\in P_N(L)} e^{-t(\underset{\sim}{r}+\underset{\sim}{n})^2+2i\underset{\sim}{k}\cdot\underset{\sim}{n}} . \tag{2.6}$$

The case $\underset{\sim}{r} = \underset{\sim}{0}$ must be considered separately. We have

$$\begin{aligned} \theta(\underset{\sim}{0},\underset{\sim}{k};P,L,N) &= \sum^{*}_{\underset{\sim}{n}\in P_N(L)} \frac{1}{|\underset{\sim}{n}|} e^{2i\underset{\sim}{k}\cdot\underset{\sim}{n}} \\ &= \lim_{\underset{\sim}{r}\to\underset{\sim}{0}} \left[\frac{\operatorname{erfc}(t|\underset{\sim}{r}|)}{|\underset{\sim}{r}|} - \frac{1}{|\underset{\sim}{r}|}\right] + \sum_{\substack{\underset{\sim}{n}\in P_N(L) \\ \underset{\sim}{n}\neq\underset{\sim}{0}}} \frac{\operatorname{erfc}(y|\underset{\sim}{n}|)}{|\underset{\sim}{n}|} \\ &\quad \times e^{2i\underset{\sim}{k}\cdot\underset{\sim}{n}} \\ &\quad + \theta_2(\underset{\sim}{0},\underset{\sim}{k};P,L,N) . \end{aligned} \tag{2.7}$$

Note that

$$\frac{\operatorname{erfc}(yx)}{x} - \frac{1}{x} = -\frac{1}{x}\operatorname{erf}(yx) \tag{2.8}$$

and

$$\operatorname{erf}(x) = \frac{2}{\sqrt{\pi}} \int_0^x e^{-y^2}\, dy = \frac{2x}{\sqrt{\pi}} - \frac{2x^3}{3\sqrt{\pi}} + \cdots \tag{2.9}$$

so that

$$\begin{aligned} \theta(\underset{\sim}{0},\underset{\sim}{k};P,L,N) &= -\frac{2y}{\sqrt{\pi}} + \sum_{\substack{\underset{\sim}{n}\in P_N(L) \\ \underset{\sim}{n}\neq\underset{\sim}{0}}} \frac{\operatorname{erfc}(y|\underset{\sim}{n}|)}{|\underset{\sim}{n}|} e^{2i\underset{\sim}{k}\cdot\underset{\sim}{n}} \\ &\quad + \theta_2(\underset{\sim}{0},\underset{\sim}{k};P,L,N) . \end{aligned} \tag{2.10}$$

We may now treat the $\underset{\sim}{r} = \underset{\sim}{0}$ case together with the $\underset{\sim}{r} \neq \underset{\sim}{0}$ case. The lattice sums in eqs. (2.10), (2.5) are absolutely convergent as $N \to \infty$ since for large $|x|$, $\operatorname{erfc} x \sim e^{-x^2}/\pi x$. Thus we may write

$$\frac{1}{2} \sum_{\underset{\sim}{n}\in P_N(L)} \sum^*_{\underset{\sim}{n}'\in P_N(L)} \sum_{i=1}^{H} \sum_{j=1}^{H} q_i q_j \, \theta_1(\underset{\sim}{r}_{ij},\underset{\sim}{k};P,L,N)$$

$$= \frac{|P|}{2|L|} N^3 \sum_{\underset{\sim}{n}\in L} \sum_{i=1}^{H} \sum_{j=1}^{H} q_i q_j \, \theta_1(\underset{\sim}{r}_{ij},\underset{\sim}{k};P,L,N) + O(N^2) \quad (2.11)$$

where $|P|$ is the volume of the region P_1. The transformation of eq. (2.2) has put all the long-ranged part of θ into the function θ_2.

To evaluate θ_2 we first note the identity [3]

$$e^{-ta^2} = \frac{1}{\sqrt{\pi t}} \int_{-\infty}^{\infty} e^{-u^2/t+2iua} \, du \, . \quad (2.12)$$

Thus, using a different component of a vector $\underset{\sim}{u}$ for each component of $\underset{\sim}{r}+\underset{\sim}{n}$, we obtain

$$e^{-t(\underset{\sim}{r}+\underset{\sim}{n})^2} = (\pi t)^{-3/2} \int d^3\underset{\sim}{u} \; e^{-\underset{\sim}{u}^2/t+2i\underset{\sim}{u}\cdot\underset{\sim}{r} + 2i\underset{\sim}{u}\cdot\underset{\sim}{n}} , \quad (2.13)$$

the integration being over all of $\underset{\sim}{u}$-space. We may now insert this into eq. (2.6) for θ_2 and obtain

$$\theta_2(\underset{\sim}{r},\underset{\sim}{k};P,L,N) = \pi^{-2} \int_0^{Y^2} dt \; t^{-2} \int d^3\underset{\sim}{u} \; e^{-\underset{\sim}{u}^2/t+2i\underset{\sim}{u}\cdot\underset{\sim}{r}} \sum_{\underset{\sim}{n}\in P_N(L)} \times e^{2i(\underset{\sim}{u}+\underset{\sim}{k})\cdot\underset{\sim}{n}} \, . \quad (2.14)$$

We may note two things about eq. (2.14). The first is that if we use the substitution $s = 1/t$, with $t^{-2}\, dt = ds$, then the integral over s may be performed immediately to give

$$\theta_2(\underset{\sim}{r},\underset{\sim}{k};P,L,N) = \pi^{-2} \int d^3\underset{\sim}{u} \; \frac{e^{-\underset{\sim}{u}^2/Y^2}}{\underset{\sim}{u}^2} \; e^{2i\underset{\sim}{u}\cdot\underset{\sim}{r}} \sum_{\underset{\sim}{n}\in P_N(L)} e^{2i(\underset{\sim}{u}+\underset{\sim}{k})\cdot\underset{\sim}{n}} \, . \quad (2.15)$$

The second is that the sum on $\underset{\sim}{n}$ is very simple. We now divide up the $\underset{\sim}{u}$-space into a sum of cells $T_{\underset{\sim}{m}}$ with centres at the points $\pi\underset{\sim}{m}$, where $\underset{\sim}{m}$ is a vector of the reciprocal lattice R. Thus

$$\underset{\sim}{u} \in T_{\underset{\sim}{m}} \quad \text{if} \quad (\underset{\sim}{u}-\pi\underset{\sim}{m}) \in V_{\pi/2} \, . \quad (2.16)$$

Eq. (2.15) may then be written

$$\theta_2(\underset{\sim}{r},\underset{\sim}{k};P,L,N) = \pi^{-2} \sum_{\underset{\sim}{m}\, R} \int_{T_{\underset{\sim}{m}}} d^3\underset{\sim}{u}\, \frac{e^{-\underset{\sim}{u}^2/y^2}}{\underset{\sim}{u}^2}\, e^{2i\underset{\sim}{u}\cdot\underset{\sim}{r}} \sum_{\underset{\sim}{n}\in P_N(L)} \times e^{2i(\underset{\sim}{u}+\underset{\sim}{k})\cdot\underset{\sim}{n}} \ . \tag{2.17}$$

If in each integral over $\underset{\sim}{u}$, we set $\underset{\sim}{u} = \pi\underset{\sim}{m}+\underset{\sim}{v}$ and use (2.16) we obtain

$$\theta_2(\underset{\sim}{r},\underset{\sim}{k};P,L,N) = \pi^{-2} \sum_{\underset{\sim}{m}\in R} e^{-\pi^2\underset{\sim}{m}^2/y^2+2\pi i\underset{\sim}{m}\cdot\underset{\sim}{r}} \int_{V_{\pi/2}} d^3\underset{\sim}{v} \times \frac{e^{-(\underset{\sim}{v}^2+2\pi\underset{\sim}{m}\cdot\underset{\sim}{v})/y^2}}{(\pi\underset{\sim}{m}+\underset{\sim}{v})^2}\, e^{2i\underset{\sim}{v}\cdot\underset{\sim}{r}} \sum_{\underset{\sim}{n}\in P_N(L)} e^{2i(\underset{\sim}{v}+\underset{\sim}{k})\cdot\underset{\sim}{n}} , \tag{2.18}$$

where we have used the result that if $\underset{\sim}{n} \in L$ and $\underset{\sim}{m} \in R$ then $\exp(2\pi i\underset{\sim}{m}\cdot\underset{\sim}{n}) = 1$. We may write eq. (2.18) in the form

$$\theta_2(\underset{\sim}{r},\underset{\sim}{k};P,L,N) = \frac{\pi}{|L|} \sum_{\underset{\sim}{m}\in R} e^{-\pi^2\underset{\sim}{m}^2/y^2+2\pi i\underset{\sim}{m}\cdot\underset{\sim}{r}}\, I_{\underset{\sim}{m}}(\underset{\sim}{r},\underset{\sim}{k};P,L,N) \tag{2.19}$$

where

$$I_{\underset{\sim}{m}}(\underset{\sim}{r},\underset{\sim}{k};P,L,N) = \sum_{\underset{\sim}{n}\in P_N(L)} e^{-2i(-\underset{\sim}{k})\cdot\underset{\sim}{n}} \left\{\frac{|L|}{\pi^3} \int_{V_{\pi/2}} d^3\underset{\sim}{v} \times \left(\frac{e^{-(\underset{\sim}{v}^2+2\pi\underset{\sim}{m}\cdot\underset{\sim}{v})/y^2}}{(\pi\underset{\sim}{m}+\underset{\sim}{v})^2}\, e^{2i\underset{\sim}{v}\cdot\underset{\sim}{r}}\right) e^{2i\underset{\sim}{v}\cdot\underset{\sim}{n}}\right\} \ . \tag{2.20}$$

If we examine eq. (2.20) for large N, in the light of the Fourier transform definitions eqs. (1.22,23), we see that eq. (2.20) is a Fourier series representation of the function

$$g_{\underset{\sim}{m}}(\underset{\sim}{r},\underset{\sim}{k}) = (\pi\underset{\sim}{m}-\underset{\sim}{k})^{-2}\, e^{-(\underset{\sim}{k}^2-2\pi\underset{\sim}{m}\cdot\underset{\sim}{k})/y^2}\, e^{-2i\underset{\sim}{k}\cdot\underset{\sim}{r}} \ . \tag{2.21}$$

The theory of Fourier series [4] shows that if $g_{\underset{\sim}{m}}(\underset{\sim}{r},\underset{\sim}{k})$ is continuous in $\underset{\sim}{k}$ then

$$I_{\underset{\sim}{m}}(\underset{\sim}{r},\underset{\sim}{k};P,L,N) = g_{\underset{\sim}{m}}(\underset{\sim}{r},\underset{\sim}{k}) + O(1/N) \ . \tag{2.22}$$

We must now note that if $\underset{\sim}{m} = \underset{\sim}{0}$ and $\underset{\sim}{k} = \underset{\sim}{0}$, $g_{\underset{\sim}{m}}(\underset{\sim}{r},\underset{\sim}{k})$ is singular in $\underset{\sim}{k}$. Thus, for $\underset{\sim}{k} \neq \underset{\sim}{0}$,

$$\theta_2(\underset{\sim}{r},\underset{\sim}{k};P,L,N) = \frac{\pi}{|L|} \sum_{\underset{\sim}{m}\in R} \frac{e^{-(\pi\underset{\sim}{m}-\underset{\sim}{k})^2/y^2}}{(\pi\underset{\sim}{m}-\underset{\sim}{k})^2} e^{2i(\pi\underset{\sim}{m}-\underset{\sim}{k})\cdot\underset{\sim}{r}} + O(1/N) \; . \tag{2.23}$$

Thus, for $\underset{\sim}{k} \neq \underset{\sim}{0}$,

$$\theta(\underset{\sim}{r},\underset{\sim}{k};P,L,N) = f(\underset{\sim}{r}) + \sum_{\substack{\underset{\sim}{n}\in L \\ \underset{\sim}{n}\neq \underset{\sim}{0}}} \frac{\operatorname{erfc}(y|\underset{\sim}{r}+\underset{\sim}{n}|)}{|\underset{\sim}{r}+\underset{\sim}{n}|} e^{2i\underset{\sim}{k}\cdot\underset{\sim}{n}}$$
$$+ \sum_{\underset{\sim}{m}\in R} \frac{e^{-(\pi\underset{\sim}{m}-\underset{\sim}{k})^2/y^2}}{|L|(\pi\underset{\sim}{m}-\underset{\sim}{k})^2} e^{2i(\pi\underset{\sim}{m}-\underset{\sim}{k})\cdot\underset{\sim}{r}} + O(1/N) \tag{2.24}$$

where

$$f(\underset{\sim}{r}) = \begin{cases} \dfrac{\operatorname{erfc}(y|\underset{\sim}{r}|)}{|\underset{\sim}{r}|} , & \underset{\sim}{r} \neq \underset{\sim}{0} \\ -2y/\sqrt{\pi} , & \underset{\sim}{r} = \underset{\sim}{0} \; . \end{cases} \tag{2.25}$$

Equation (2.24) gives a representation of $\theta(r,k;P,L,N)$ in terms of two lattice sums which are absolutely convergent. For $\underset{\sim}{k} \neq \underset{\sim}{0}$, the sums are uniformly convergent in $\underset{\sim}{r}$ for all $\underset{\sim}{r}$ and uniformly convergent in $\underset{\sim}{k}$ in any closed subregion of $V_{\pi/2}$ which does not include $\underset{\sim}{k} = \underset{\sim}{0}$. This means that gradients of $\theta(\underset{\sim}{r},\underset{\sim}{k};P,L,N)$ with respect to $\underset{\sim}{r}$ may be evaluated using term-by-term differentiation of the series in eq. (2.24). Notice that for $\underset{\sim}{k}$ small, $\theta(\underset{\sim}{r},\underset{\sim}{k};P,L,N$ contains a singular term which is the $\underset{\sim}{m} = \underset{\sim}{0}$ term in the sum over $\underset{\sim}{m} \in R$, namely

$$\frac{\pi}{|L|} e^{-\underset{\sim}{k}^2/y^2} \frac{e^{-2i\underset{\sim}{k}\cdot\underset{\sim}{r}}}{\underset{\sim}{k}^2} \simeq \frac{\pi}{|L|} e^{-\underset{\sim}{k}^2/y^2} \left[\frac{1}{\underset{\sim}{k}^2} - \frac{2i\underset{\sim}{k}\cdot\underset{\sim}{r}}{\underset{\sim}{k}^2} - \frac{2(\underset{\sim}{k}\cdot\underset{\sim}{r})^2}{\underset{\sim}{k}^2} + \cdots\right]. \tag{2.26}$$

The most singular term in eq. (2.26) diverges as $\underset{\sim}{k}^{-2}$ for small $\underset{\sim}{k}$, but is independent of $\underset{\sim}{r}$. Thus the sum

$$\sum_{i=1}^{H} \sum_{j=1}^{H} q_i q_j \, \theta(\underset{\sim}{r}_{ij},\underset{\sim}{k};P,L,N) \tag{2.27}$$

will not contain this divergence because of the charge neutrality condition

$$\sum_{i=1}^{H} q_i = 0 \ . \tag{2.28}$$

The next most singular term in eq. (2.26) will contribute

$$-\frac{2\pi i}{|L|k^2} e^{-k^2/y^2} \sum_{i=1}^{H} \sum_{j=1}^{H} q_i q_j (\underset{\sim}{r}_i - \underset{\sim}{r}_j)\cdot\underset{\sim}{k}$$

to the sum in (2.27) and this contribution is again zero by eq. (2.28). The final singular term in eq. (2.26) makes the contribution

$$-\frac{2\pi}{|L|} \frac{e^{-\underset{\sim}{k}^2/y^2}}{\underset{\sim}{k}^2} \sum_{i=1}^{H} \sum_{j=1}^{H} q_i q_j [\underset{\sim}{k}\cdot(\underset{\sim}{r}_i - \underset{\sim}{r}_j)]^2 \tag{2.29}$$

to the sum in eq. (2.27). We may expand out the square in (2.29) as

$$[\underset{\sim}{k}\cdot(\underset{\sim}{r}_i - \underset{\sim}{r}_j)]^2 = (\underset{\sim}{k}\cdot\underset{\sim}{r}_i)^2 - 2(\underset{\sim}{k}\cdot\underset{\sim}{r}_i)(\underset{\sim}{k}\cdot\underset{\sim}{r}_j) + (\underset{\sim}{k}\cdot\underset{\sim}{r}_j)^2$$

and use the charge neutrality condition and

$$\underset{\sim}{M} = \sum_{i=1}^{H} q_i \underset{\sim}{r}_i \ . \tag{2.30}$$

This shows that the $\underset{\sim}{m} = \underset{\sim}{0}$ terms in the sum (2.27) contains the singular contribution

$$-\frac{4\pi}{|L|} e^{-\underset{\sim}{k}^2/y^2} \frac{(\underset{\sim}{M}\cdot\underset{\sim}{k})^2}{\underset{\sim}{k}^2} \tag{2.31}$$

This term is singular as $\underset{\sim}{k} \to \underset{\sim}{0}$ since although it is finite for all $\underset{\sim}{k}$, it does not have a unique limit as $\underset{\sim}{k} \to \underset{\sim}{0}$. As we shall see below, at $\underset{\sim}{k} = 0$ this singularity makes a shape dependent contribution to the electrostatic energy of an ionic crystal when $\underset{\sim}{M} \neq \underset{\sim}{0}$.

We must now return to the evaluation of

$$\theta(\underset{\sim}{r},\underset{\sim}{0};P,L,N) = \sum_{\underset{\sim}{n}\in P_N(L)} \frac{1}{|\underset{\sim}{r}+\underset{\sim}{n}|} \ . \tag{2.32}$$

This evaluation proceeds exactly as for $\underset{\sim}{k} \neq \underset{\sim}{0}$ except for the evaluation of $I_{\underset{\sim}{0}}(\underset{\sim}{r},\underset{\sim}{0};P,L,N)$ in eq. (2.19). Thus we obtain

$$\theta(\underset{\sim}{r},\underset{\sim}{0};P,L,N) = f(\underset{\sim}{r}) + \sum_{\substack{\underset{\sim}{n}\in L \\ \underset{\sim}{n}\neq\underset{\sim}{0}}} \frac{\operatorname{erfc}(y|\underset{\sim}{r}+\underset{\sim}{n}|)}{|\underset{\sim}{r}+\underset{\sim}{n}|} + \sum_{\substack{\underset{\sim}{m}\in R \\ \underset{\sim}{m}\neq\underset{\sim}{0}}} \frac{e^{-\pi^2 \underset{\sim}{m}^2/y^2}}{\pi|L|\underset{\sim}{m}^2} e^{2\pi i \underset{\sim}{m}\cdot\underset{\sim}{r}} + \frac{\pi}{|L|} \underset{\sim}{I_0}(\underset{\sim}{r},\underset{\sim}{0};P,L,N) + O(1/N) , \tag{2.33}$$

with

$$\underset{\sim}{I_0}(\underset{\sim}{r},\underset{\sim}{0};P,L,N) = \frac{|L|}{\pi^3} \int_{V_{\pi/2}} d^3\underset{\sim}{v} \frac{e^{-\underset{\sim}{v}^2/y^2+2i\underset{\sim}{v}\cdot\underset{\sim}{r}}}{\underset{\sim}{v}^2} \sum_{\underset{\sim}{n}\in P_N(L)} e^{2i\underset{\sim}{v}\cdot\underset{\sim}{n}} . \tag{2.34}$$

We may now note that if $P_N(L)$ is symmetric in the sense that if $\underset{\sim}{n} \in P_N(L)$ then $-\underset{\sim}{n} \in P_N(L)$, the only odd part of the integral in (2.34) is the odd part of $\exp(2i\underset{\sim}{v}\cdot\underset{\sim}{r})$, namely $i \sin(2i\underset{\sim}{v}\cdot\underset{\sim}{r})$. Thus we may replace this exponential by $\cos 2\underset{\sim}{v}\cdot\underset{\sim}{r} = 1 - 2 \sin^2(\underset{\sim}{v}\cdot\underset{\sim}{r})$ and obtain

$$\underset{\sim}{I_0}(\underset{\sim}{r},\underset{\sim}{0};P,L,N) = \underset{\sim}{I_0}(\underset{\sim}{0},\underset{\sim}{0};P,L,N) - 2\frac{|L|}{\pi^3} \int_{V_{\pi/2}} d^3\underset{\sim}{v}\, e^{-\underset{\sim}{v}^2/y^2} \frac{\sin^2(\underset{\sim}{v}\cdot\underset{\sim}{r})}{\underset{\sim}{v}^2} \sum_{\underset{\sim}{n}\in P_N(L)} \times e^{2i\underset{\sim}{v}\cdot\underset{\sim}{n}} . \tag{2.35}$$

The function $I_0(0,0;P,L,N)$ is in fact $O(N^2)$, and contains the obvious divergence as $\tilde{N} \tilde{\to} \tilde{\infty}$ in eq. (2.32). However, it is independent of $\underset{\sim}{r}$ and so this divergence vanishes by charge neutrality when we use $\theta(\underset{\sim}{r},\underset{\sim}{0};P,L,N)$ in calculating energies of ionic crystals. Thus we need not consider it further here.

We may now turn to the sum on $n \in P_N(L)$ in eq. (2.35). This sum is $O(N^3)$ if $v = 0$ but $O(1)$ for $\tilde{v} \neq \tilde{0}$. Thus the integral in (2.35) is dominated by the $\underset{\sim}{v} \approx \underset{\sim}{0}$ behaviour of the integrand. Thus we set $\underset{\sim}{w} = N\underset{\sim}{v}$ and obtain

$$\underset{\sim}{I_0}(\underset{\sim}{r},\underset{\sim}{0};P,L,N) = \underset{\sim}{I_0}(\underset{\sim}{0},\underset{\sim}{0};P,L,N) - J(\underset{\sim}{r};P,L,N$$

with

$$J(\underset{\sim}{r};P,L,N = 2\pi^{-3} \int_{V_{N\pi/2}} d^3\underset{\sim}{w}\, e^{-\underset{\sim}{w}^2/y^2N^2} \frac{N^2\sin^2(\underset{\sim}{w}\cdot\underset{\sim}{r}/N)}{\underset{\sim}{w}^2} \times \sum_{\underset{\sim}{n}\in P_N(L)} e^{2i\underset{\sim}{w}\cdot\underset{\sim}{n}/N} \frac{|L|}{N^3} \tag{2.36}$$

The finite sum on $\underset{\sim}{n}$ in eq. (2.36) is a finite sum approximation to the Riemann integral

$$\int_{P_1} d^3\underset{\sim}{x}\; e^{2i\underset{\sim}{w}\cdot\underset{\sim}{x}} , \tag{2.37}$$

the corrections being O(1/N) for large N. We may insert (2.37) for the sum in eq. (2.36) and expand the integrand in inverse powers of N. We obtain

$$J(\underset{\sim}{r};P,L,N) = 2\pi^{-3}\int d^3\underset{\sim}{w}\,\frac{(\underset{\sim}{r}\cdot\underset{\sim}{w})^2}{\underset{\sim}{w}^2}\int_{P_1} e^{2i\underset{\sim}{w}\cdot\underset{\sim}{x}}\,d^3\underset{\sim}{x} + O(1/N)$$

$$\equiv j(\underset{\sim}{r},P) + O(1/N) . \tag{2.38}$$

Thus

$$I_{\underset{\sim}{0}}(\underset{\sim}{r},\underset{\sim}{0};P,L,N) = I_{\underset{\sim}{0}}(\underset{\sim}{0},\underset{\sim}{0};P,L,N) - j(\underset{\sim}{r},P) + O(1/N) \tag{2.39}$$

and so, from eq. (2.33),

$$\theta(\underset{\sim}{r},\underset{\sim}{0};P,L,N) = \frac{\pi}{|L|}\,I_{\underset{\sim}{0}}(\underset{\sim}{0},\underset{\sim}{0};P,L,N) + \psi_E(\underset{\sim}{r},L) - \frac{\pi}{|L|}\times j(\underset{\sim}{r},P) + O(1/N) \tag{2.40}$$

where

$$\psi_E(\underset{\sim}{r},L) = f(\underset{\sim}{r}) + \sum_{\substack{\underset{\sim}{n}\in L\\ \underset{\sim}{n}\neq \underset{\sim}{0}}}\frac{\mathrm{erfc}(y|\underset{\sim}{r}+\underset{\sim}{n}|)}{|\underset{\sim}{r}+\underset{\sim}{n}|} + \sum_{\substack{\underset{\sim}{m}\in R\\ \underset{\sim}{m}\neq \underset{\sim}{0}}}\frac{e^{-\pi^2\underset{\sim}{m}^2/y^2}}{\pi|L|\underset{\sim}{m}^2}\,e^{2\pi i\underset{\sim}{m}\cdot\underset{\sim}{r}} . \tag{2.41}$$

We wish to consider two cases: P_1 a sphere of radius 1 and P_1 a plane slab of width 2.

(a) P_1 a sphere of radius 1.

For this case

$$j(\underset{\sim}{r};P) = 2\pi^{-3}\int d^3\underset{\sim}{w}\,\frac{(\underset{\sim}{r}\cdot\underset{\sim}{w})^2}{\underset{\sim}{w}^2}\int_{P_1} e^{2i\underset{\sim}{w}\cdot\underset{\sim}{x}}\,d^2\underset{\sim}{x}$$

$$= \frac{2\pi^{-3}}{3}\sum_{i=1}^{3}\underset{\sim}{r}_i^{\,2}\int_{P_1} d^3\underset{\sim}{x}\int d^3\underset{\sim}{w}\,e^{2i\underset{\sim}{w}\cdot\underset{\sim}{x}} \tag{2.42}$$

where we have used the symmetry of P_1 with respect to the three components of $\underset{\sim}{x}$. The integral over $\underset{\sim}{w}$ in the second form of (2.42) is $\pi^3\delta(\underset{\sim}{r})$ so that for the case when P is a sphere of radius 1,

$$j(\underset{\sim}{r},P) = \frac{2\underset{\sim}{r}^2}{3} . \tag{2.43}$$

(b) P_1 a plane slab of width 2 normal to a z-axis.
For this case

$$\int_{P_1} e^{2i\underset{\sim}{w}\cdot\underset{\sim}{x}}\, d^3\underset{\sim}{x} = \pi^2\delta(w_1)\delta(w_2)\frac{1}{w_3}\ \sin 2w_3 \tag{2.44}$$

so that

$$j(\underset{\sim}{r},P) = 2\pi^{-2}\ \underset{\sim}{r}_3^{\,2}\int_{-\infty}^{\infty} dw_3\ \frac{1}{w_3}\sin 2w_3 = 2\underset{\sim}{r}_3^{\,2}\ . \tag{2.45}$$

We concluce that for P a sphere

$$\theta(\underset{\sim}{r},\underset{\sim}{0};P,L,N) = \frac{\pi}{|L|}\ I_{\underset{\sim}{0}}(\underset{\sim}{0},\underset{\sim}{0};P,L,N) + \psi_E(\underset{\sim}{r},L) - \frac{2\pi}{3|L|}\ \underset{\sim}{r}^2 + O(1/N) \tag{2.46}$$

while for P a plane slab normal to the z-direction

$$\theta(\underset{\sim}{r},\underset{\sim}{0};P,L,N) = \frac{\pi}{|L|}\ I_{\underset{\sim}{0}}(\underset{\sim}{0},\underset{\sim}{0};P,L,N) + \psi_E(\underset{\sim}{r},L) - \frac{2\pi}{|L|}\ \underset{\sim}{r}_3^{\,2} + O(1/N)\ . \tag{2.47}$$

These results complete the discussion of the function $\theta(\underset{\sim}{r},\underset{\sim}{k};P,L,N)$ defined in eq. (1.29).

III. ELECTROSTATIC ENERGY OF IONIC CRYSTALS

First we consider the electrostatic energy of a unit cell in the centre of a P-shaped piece of ionic crystal. We may write this in the form

$$\begin{aligned} e_N(L,P) &= \frac{1}{2}\sum_{\underset{\sim}{n}\in P_N(L)}{}^{*}\ \sum_{i=1}^{H}\sum_{j=1}^{H} q_i q_j |\underset{\sim}{r}_{ij}+\underset{\sim}{n}|^{-1} \\ &= \frac{1}{2}\sum_{i=1}^{H}\sum_{j=1}^{H} q_i q_j\ \theta(\underset{\sim}{r}_{ij},\underset{\sim}{0};P,L,N)\ . \end{aligned} \tag{3.1}$$

We may now insert the representation eq. (2.40) for θ. Further, we may note that

$$-\frac{1}{2}\frac{\pi}{|L|}\sum_{i=1}^{H}\sum_{j=1}^{H} q_i q_j\ j(\underset{\sim}{r}_{ij},P) = \frac{\pi}{|L|}\ j(\underset{\sim}{M},P) \tag{3.2}$$

where $\underset{\sim}{M}$ is given in eqs. (1.11) or (2.30). This comes from the definition of $j(\underset{\sim}{r},P)$ in eq. (2.38) and the result

$$\sum_{i=1}^{H}\sum_{j=1}^{H} q_i q_j(\underset{\sim}{r}_{ij}\cdot\underset{\sim}{v})^2 = -2\sum_{i=1}^{H}\sum_{j=1}^{H} q_i q_j\ \underset{\sim}{r}_i\cdot\underset{\sim}{r}_j = -2\underset{\sim}{M}^2\ . \tag{3.3}$$

Thus

$$e_N(L,P) = \frac{1}{2}\sum_{i=1}^{H}\sum_{j=1}^{H} q_i q_j \psi_E(\underset{\sim}{r}_{ij},L) + \frac{\pi}{|L|} j(\underset{\sim}{M},P) + O(1/N) . \tag{3.4}$$

The coefficient of $I_0(\underset{\sim}{0},\underset{\sim}{0};P,L,N)$ in eq. (3.4) is the square of the net charge in a unit cell and thus zero. For the case when P is a sphere

$$e_N(L,P) = \frac{1}{2}\sum_{i=1}^{H}\sum_{j=1}^{H} q_i q_j \psi_E(\underset{\sim}{r}_{ij},L) + \frac{2\pi}{3|L|} \underset{\sim}{M}^2 + O(1/N) \tag{3.5}$$

and for the case when P is a plane slab normal to the z-direction,

$$e_N(L,P) = \frac{1}{2}\sum_{i=1}^{H}\sum_{j=1}^{H} q_i q_j \psi_E(\underset{\sim}{r}_{ij},L) + \frac{2\pi}{|L|} M_z^{\,2} + O(1/N) . \tag{3.6}$$

Very similar techniques to those used in section 2 for $\theta(\underset{\sim}{r},\underset{\sim}{0};P,L,N)$ may be used to evaluate the total electrostatic energy [5]

$$E_N(L,P) = \frac{1}{2} \sum_{\underset{\sim}{n}\in P_N(L)} \; \sum^{*}_{\underset{\sim}{n}'\in P_N(L)} \; \sum_{i=1}^{H}\sum_{j=1}^{H} q_i q_j |\underset{\sim}{r}_{ij}+\underset{\sim}{n}|^{-1} \tag{3.7}$$

of a piece P_N of ionic crystal of identical unit cells. The total number of unit cells in P_N is $N^3|P|/|L|$, when $|P|$ is finite (which it isn't for a plane slab). The detailed analysis gives

$$E_N(L,P) = [N^3|P|/|L|]e_N(L,P) + O(N^2) . \tag{3.8}$$

We now turn to the examples of cesium chloride and fluorite structures discussed in section 1. The energy per unit cell in the limit of a large piece of crystal ($N \to \infty$) is shape dependent:

$$e(L,P) = \frac{1}{2}\sum_{i=1}^{H}\sum_{j=1}^{H} q_i q_j \psi_E(\underset{\sim}{r}_{ij},L) + \frac{\pi}{|L|} j(\underset{\sim}{M},P) . \tag{3.9}$$

Notice that if $\underset{\sim}{M} = \underset{\sim}{0}$

$$j(\underset{\sim}{M},P) = 0 . \tag{3.10}$$

If we consider example (i), eq. (1.16 et seq.) of a CsCl structure, then the energy per unit cell is, for case (a) with $\underset{\sim}{M} = \underset{\sim}{0}$,

$$e_a(L) = q^2[\psi_E(0,L) - \psi_E((\tfrac{1}{2},\tfrac{1}{2},\tfrac{1}{2}),L)] \tag{3.11}$$

while for case (b), with $\underset{\sim}{M} = \frac{1}{2}q(1,1,1)$, it is

$$e_b(L) = e_a(L) + \pi q^2/2|L| \tag{3.12}$$

for both the plane slab and spherical cases. For the fluorite structure,

$$e_b(L) = e_a(L) + 8\pi q^2/|L| \tag{3.13}$$

for both the plane slab and spherical cases. The only difference between the two cases is that the origin of the unit cell has been chosen differently in cases (a) and (b). We must now sort out why the choice of origin of the unit cell affects the energy per unit cell in the limit of a large crystal. The piece P_N of ionic crystal is composed of identical unit cells. Thus the difference between the two cases (a) and (b) is that their surface structures are different. This means that the energy per unit cell of an ionic crystal in the limit as the crystal becomes large, is affected by the structure on the surface of the piece of crystal. The normal assumption that surface contributions tend to zero when the surface to volume ratio tends to zero does not apply.

We may consider case (a) of the CsCl lattice in plane slab geometry. There is one surface at z = -NB and one at z = NB where B is the lattice spacing and the slab is 2N unit cells thick. These surfaces have half positive charges on them. To convert this slab into a slab of case (b) CsCl structure, we must put a layer of half positive charges (charge density $q/2B^2$) on the surface at z = NB and a layer of half positive and a layer of full negative charges (net charge density $-q/2B^2$) on the surface at z = -NB. If we treat this system as a condenser and set the potential at z = -NB to be zero, then the potential at z = NB is O(NB). Thus the charges at z = NB have an energy of interaction with those at z = -NB which is O(N) and this energy makes an O(1) contribution to the energy per unit cell in the limit $N \to \infty$. These considerations show that a change in the surface structure of an ionic crystal can indeed change the electrostatic energy per unit cell in the limit of a large piece of crystal.

In general the electrostatic energy per unit cell of an infinite P-shaped piece of ionic crystal may be written

$$e(L,P) = \frac{1}{2}\sum_{i=1}^{H}\sum_{j=1}^{H} q_i q_j\, \psi_E(\underset{\sim}{r}_{ij},L) + \sum_{i=1}^{3} B_i(P)M_i^2 \,. \tag{3.14}$$

We may consider the effects of attaching to the surfaces of a cube of ionic crystal, excess charge layers. If we attach layers of evenly distributed charges of charge density σ_i to the positive i surface, and $-\sigma_i$ to the negative i surface, then detailed calculation gives

$$e(L,P) = \frac{1}{2} \sum_{i=1}^{H} \sum_{j=1}^{H} q_i q_j \psi_E(\underset{\sim}{r}_{ij},L) + \frac{2\pi}{3|L|} \sum_{i=1}^{3} (M_i + L^3\sigma_i)^2 . \tag{3.15}$$

This energy per unit cell may be minimized with respect to the σ_i and the minimum is

$$e(L,P) = \frac{1}{2} \sum_{i=1}^{H} \sum_{j=1}^{H} q_i q_j \psi_E(\underset{\sim}{r}_{ij},L) , \tag{3.16}$$

the standard Madelung result. This suggests that when a crystal shape is produced either by growing a crystal or fracturing a larger crystal, the resulting crystal is not composed of identical unit cells. The surface structure adapts in such a way that the minimization of electrostatic energy per unit cell discussed here occurs and the energy per unit cell is given by the standard Madelung result.

IV. DIPOLAR LATTICE SUMS

One of the problems introduced in section 1 of this lecture is the evaluation of

$$T(\underset{\sim}{k},\underset{\sim}{r}) = \sum_{\underset{\sim}{m}\in P_N(L)} \nabla\nabla|\underset{\sim}{r}+\underset{\sim}{m}|^{-1} e^{2i\underset{\sim}{k}\cdot\underset{\sim}{m}} \tag{4.1}$$

and especially its evaluation at $\underset{\sim}{k} = \underset{\sim}{0}$. Since the sum in (4.1) is finite, we may write

$$\begin{aligned} T(\underset{\sim}{k},\underset{\sim}{r}) &= \nabla\nabla\left\{ \sum_{\underset{\sim}{m}\in P_N(L)} |\underset{\sim}{r}+\underset{\sim}{m}|^{-1} e^{2i\underset{\sim}{k}\cdot\underset{\sim}{m}} \right\} \\ &= \nabla\nabla\, \theta(\underset{\sim}{r},\underset{\sim}{k};P,L,N) , \end{aligned} \tag{4.2}$$

the gradients being with respect to $\underset{\sim}{r}$. For $\underset{\sim}{k} \neq \underset{\sim}{0}$,

$$\theta(\underset{\sim}{r},\underset{\sim}{k};P,L,N) = \Psi_E(\underset{\sim}{r},\underset{\sim}{k},L) + \frac{\pi}{|L|\underset{\sim}{k}^2} e^{-\underset{\sim}{k}^2/y^2} e^{-2i\underset{\sim}{k}\cdot\underset{\sim}{r}} + O(1/N) ,$$

where

$$\Psi_E(\underset{\sim}{r},\underset{\sim}{k},L) = f(\underset{\sim}{r}) + \sum_{\substack{\underset{\sim}{n}\in L\\ \underset{\sim}{n}\neq \underset{\sim}{0}}} \frac{\text{erfc}(y|\underset{\sim}{r}+\underset{\sim}{n}|)}{|\underset{\sim}{r}+\underset{\sim}{n}|} e^{2i\underset{\sim}{k}\cdot\underset{\sim}{n}} + \sum_{\substack{\underset{\sim}{m}\in R\\ \underset{\sim}{m}\neq \underset{\sim}{0}}} \frac{\pi e^{-(\pi\underset{\sim}{m}-\underset{\sim}{k})^2/y^2}}{|L|(\pi\underset{\sim}{m}-\underset{\sim}{k})^2} e^{2i(\pi\underset{\sim}{m}-\underset{\sim}{k})\cdot\underset{\sim}{r}} . \tag{4.4}$$

The function $\Psi_E(\underset{\sim}{r},\underset{\sim}{k},L)$ contains two lattice sums which are absolutely convergent and uniformly convergent in $\underset{\sim}{r}$. Thus they may be differentiated term by term. This gives

$$T(\underset{\sim}{k},\underset{\sim}{r}) = \nabla\nabla\, \Psi_E(\underset{\sim}{r},\underset{\sim}{k},L) - 4\pi \frac{\underset{\sim}{k}\underset{\sim}{k}}{\underset{\sim}{k}^2} e^{-\underset{\sim}{k}^2/y^2 - 2i\underset{\sim}{k}\cdot\underset{\sim}{r}} + O(1/N) . \tag{4.5}$$

For $\underset{\sim}{k} = \underset{\sim}{0}$,

$$\theta(\underset{\sim}{r},\underset{\sim}{0};P,L,N) = I_{\underset{\sim}{0}}(\underset{\sim}{0},\underset{\sim}{0};P,L,N) + \psi_E(\underset{\sim}{r},L) - \frac{\pi}{|L|} j(\underset{\sim}{r},P) + O(1/N) . \tag{4.6}$$

Thus

$$T(\underset{\sim}{0},\underset{\sim}{r}) = \nabla\nabla\, \psi_E(\underset{\sim}{r},L) - \frac{\pi}{|L|} \nabla\nabla\, j(\underset{\sim}{r},P) + O(1/N) . \tag{4.7}$$

We may write

$$j(\underset{\sim}{r},P) = \sum_{i=1}^{3} r_i^{\,2} B_i(P) \tag{4.8}$$

where

$$B_j(P) = 2\pi^{-3} \int d^3\underset{\sim}{w}\, \frac{w_j^{\,2}}{\underset{\sim}{w}^2} \int_{P_1} e^{2i\underset{\sim}{w}\cdot\underset{\sim}{x}}\, d^3\underset{\sim}{x} . \tag{4.9}$$

We may now calculate

$$\nabla\nabla\, j(\underset{\sim}{r},P)_{ij} = 2\, B_i(P)\delta_{ij} \tag{4.10}$$

and so

$$T(\underset{\sim}{0},\underset{\sim}{r})_{ij} = \nabla\nabla\, \psi_E(\underset{\sim}{r},L)_{ij} - \frac{2\pi}{|L|} B_i(P)\delta_{ij} . \tag{4.11}$$

To complete this section we give $\nabla\nabla\, \Psi_E(\underset{\sim}{r},\underset{\sim}{k},L)$ and $\nabla\nabla\, \psi_E(\underset{\sim}{r},L)$ in explicit form. For $\underset{\sim}{r} \neq \underset{\sim}{0}$,

$$\nabla\nabla\,\psi_E(\mathbf{r},L) = \sum_{\mathbf{n}\in L}\left\{\left[\frac{3(\mathbf{r}+\mathbf{n})(\mathbf{r}+\mathbf{n})}{|\mathbf{r}+\mathbf{n}|^2} - I\right]\left[\frac{\operatorname{erfc}(y|\mathbf{r}+\mathbf{n}|)}{|\mathbf{r}+\mathbf{n}|^3} + \frac{2y}{\sqrt{\pi}\,|\mathbf{r}+\mathbf{n}|^2}\right.\right.$$
$$\left.\times\, e^{-y^2(\mathbf{r}+\mathbf{n})^2}\right]$$
$$\left.+\,\frac{4y^3}{\sqrt{\pi}}\,\frac{(\mathbf{r}+\mathbf{n})(\mathbf{r}+\mathbf{n})}{|\mathbf{r}+\mathbf{n}|^2}\,e^{-y^2(\mathbf{r}+\mathbf{n})^2}\right\}$$
$$-\,\frac{4\pi}{|L|}\sum_{\substack{\mathbf{m}\in R\\ \mathbf{m}\neq\mathbf{0}}}\frac{\mathbf{m}\mathbf{m}}{|\mathbf{m}|^2}\,e^{-\pi^2\mathbf{m}^2/y^2+2\pi i\mathbf{m}\cdot\mathbf{r}} \tag{4.12}$$

and

$$\nabla\nabla\,\Psi_E(\mathbf{r},\mathbf{k},L) = \sum_{\mathbf{n}\in L}\left\{\left[\frac{3(\mathbf{r}+\mathbf{n})(\mathbf{r}+\mathbf{n})}{|\mathbf{r}+\mathbf{n}|^2} - I\right]\left[\frac{\operatorname{erfc}(y|\mathbf{r}+\mathbf{n}|)}{|\mathbf{r}+\mathbf{n}|^3} + \frac{2y}{\sqrt{\pi}}\right.\right.$$
$$\left.\times\,\frac{e^{-y^2(\mathbf{r}+\mathbf{n})^2}}{|\mathbf{r}+\mathbf{n}|^2}\right]$$
$$\left.+\,\frac{4y^2}{\sqrt{\pi}}\,\frac{(\mathbf{r}+\mathbf{n})(\mathbf{r}+\mathbf{n})}{|\mathbf{r}+\mathbf{n}|^2}\,e^{-y^2(\mathbf{r}+\mathbf{n})^2}\right\}e^{2i\mathbf{k}\cdot\mathbf{n}}$$
$$-\,\frac{4\pi}{|L|}\sum_{\substack{\mathbf{m}\in R\\ \mathbf{m}\neq\mathbf{0}}}\frac{(\pi\mathbf{m}-\mathbf{k})(\pi\mathbf{m}-\mathbf{k})}{(\pi\mathbf{m}-\mathbf{k})^2}\,e^{-(\pi\mathbf{m}-\mathbf{k})^2/y^2}$$
$$\times\, e^{2i(\pi\mathbf{m}-\mathbf{k})\cdot\mathbf{r}}\ . \tag{4.13}$$

To evaluate these objects for $\mathbf{r} = \mathbf{0}$ note that $\theta(\mathbf{0},\mathbf{k};P,L,N)$ is defined as

$$\lim_{\mathbf{r}\to\mathbf{0}}\left[\theta(\mathbf{r},\mathbf{k};P,L,N) - \frac{1}{|\mathbf{r}|}\right]\ . \tag{4.14}$$

Thus for $\mathbf{r} = \mathbf{0}$, eqs. (4.12 and 13) must be corrected by replacing the $\mathbf{n} = \mathbf{0}$ term in the sum on $\mathbf{n} \in L$ by

$$\lim_{\mathbf{r}\to\mathbf{0}}\nabla\nabla\left(\frac{\operatorname{erfc}(y|\mathbf{r}|)}{|\mathbf{r}|} - \frac{1}{|\mathbf{r}|}\right) = -\lim_{\mathbf{r}\to\mathbf{0}}\frac{\operatorname{erf}(y|\mathbf{r}|)}{|\mathbf{r}|}\ . \tag{4.15}$$

We may use the series expansion

$$\mathrm{erf}(x) = \frac{2}{\sqrt{\pi}}\left(x - \frac{x^3}{3} + \frac{x^5}{10} + \cdots\right) \tag{4.16}$$

to evaluate eq. (4.15). For $\underset{\sim}{r} = \underset{\sim}{0}$, the $\underset{\sim}{n} = \underset{\sim}{0}$ terms in the sums on $\underset{\sim}{n} \in L$ in eqs. (4.12, 13) must be replaced by

$$\frac{4y^3}{3\sqrt{\pi}}\, I\ . \tag{4.17}$$

V. CONCLUSIONS

We have seen that the electrostatic energy per unit cell of a piece of ionic crystal of shape P made up of identical unit cells is, in the limit when the piece of crystal becomes large, dependent on the shape P of the crystal, if the unit cell of the crystal has a net dipole moment. Whether a unit cell has a dipole moment or not depends on the choice of origin for the unit cell. This dependence on choice of origin is resolved by studying the surface of crystals made up of unit cells with different origins. Real crystals appear not to be made up of identical unit cells. A surface of a crystal which would be polar if the crystal were made up only of identical unit cells in fact adjusts its surface structure in such a way that the shape dependent terms in the unit cell electrostatic energy disappear.

The lattice sums which we have been discussing are sometimes conditionally convergent and sometimes, as in the case of $\theta(\underset{\sim}{r},\underset{\sim}{k};P,L,N)$ divergent as the lattice becomes large. The object of the study of these sums has not been to develop simple analytic forms but to obtain absolutely (and rapidly) convergent sum representations for the original sums. While these representations can become quite complicated, they are all, in fact, fairly rapidly calculated. In one of his lectures, Rahman discusses the problem of producing a very fast algorithm for the evaluation of these absolutely convergent lattice sums. This algorithm becomes necessary when the sums must be evaluated a very large number of times, as in a simulation of an ionic or dipolar system in periodic boundary conditions.

The equations for the distortion of an ionic crystal due to an external field or a set of defects also contain shape dependent lattice sums. These have been calculated explicitly. A further application of these last sums in molecular dynamics and Monte Carlo simulations of ionic and dipolar systems will be discussed in a later lecture.

REFERENCES

1. De Leeuw, S.W., Perram, J.W. and Smith, E.R., Proc. Roy. Soc. (Lond.) A. In press.
2. Whittaker, E.T. and Watson, G.N., "Modern Analysis", (1958) ch. 12. Cambridge University Press.
3. Gradshteyn, I-S. and Ryzhik, I.M., "Table of Integrals, Series and Products", (1965) p. 480. Academic Press.
4. See ref. 2, ch. 10.
5. Smith, E.R., Proc. Roy. Soc. (Lond.) A. In press.

ON THE CRYSTALLINE, LIQUID, GLASSY, GASEOUS, AND SUPERFLUID STATES OF SIMPLE SUBSTANCES

R. M. J. Cotterill

Department of Structural Properties of Materials
The Technical University of Denmark
Building 307, DK-2800 Lyngby, Denmark

1. INTRODUCTION

The appearance of this communication, which does not specifically deal with fast ionic conduction, in a volume devoted to that subject, obviously requires prefatory comment. Its inclusion seems justified on at least two counts. Explanations of the superionic conduction phenomenon are frequently couched in terms of the melting of a sublattice. An understanding of the transitions between the fundamental forms of simple matter might thus be a prerequisite for the proper appreciation of the superionic state. Moreover, it has recently been found that the glassy versions of certain substances display superionic conduction [1]. It seems unlikely that a satisfactory explanation of this intriguing fact will actually precede the emergence of an acceptable theory of the glassy state itself.

The objective of this brief review is to provide a commentary on some current developments in the quest to explain the different states of simple matter within the framework of a single approach. Among recent surveys which have covered certain aspects of the same issue, Anderson's review [2] is particularly useful in that it summarizes the arguments for and against the two most prominent glass models: the highly dislocated crystal, and what might be called a fundamental glassy state. The present article first touches briefly on some recent evidence against dislocation-mediated melting, in three dimensions, and then discusses the free volume and bifurcation approaches to the glass-liquid and crystal-liquid transitions. Contact is then made with the interesting limiting case of the hard sphere model, and the penultimate section includes a few qualitative remarks on the bifurcation theory of superfluidity.

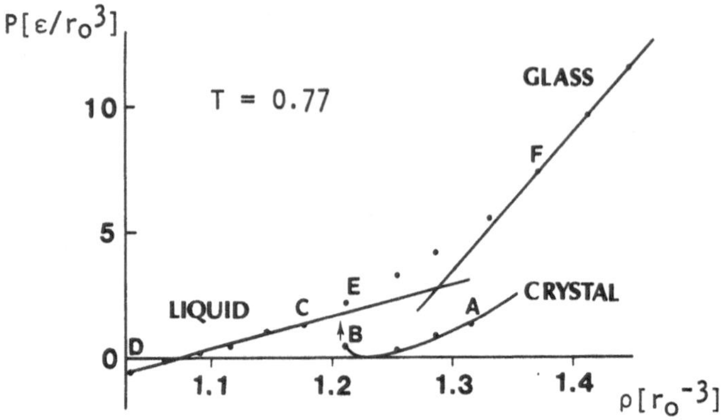

Figure 1. The T = 0.77 isotherms of a crystal, liquid, and pressure-induced glass, with computed points indicated [14]. The arrow indicates the position of crystal instability, and the intersection of the extrapolated lines locates a formal glass transition density. The letters A through F indicate the expansion and compression sequence of states. The calculations were made on the basis of the diatomic Lennard-Jones interaction energy, $V_{LJ} = \varepsilon((r_o/r)^{12} - 2(r_o/r)^6)$; reduced units were used throughout, i.e. lengths in units of r_o, energies ε, temperatures ε/k_B, densities r_o^{-3}, pressures ε/r_o^3. k_B is Boltzmann's constant.

2. EVIDENCE AGAINST A CRYSTAL INSTABILITY IN MELTING

The dislocation theory of melting, from its implicit emergence in an early paper by Mott and Gurney [3], through a gradual development by Bragg [4], Shockley [5], Ookawa [6], Mizushima [7], and Kuhlmann-Wilsdorf [8], was the subject of a recent review [9]. The later approach to the melting of two-dimensional crystals, in terms of dislocation dipole generation, first observed in a computer simulation [10] and shortly thereafter embodied in the now well-known theory of Kosterlitz and Thouless [11], is a separate story. Although the dipole pair of two-dimensional theory has an obvious counterpart in three dimensions, namely the dislocation loop, and although such loops have been observed in three-dimensional melting, as studied by computer simulation [12], caution should be applied to the question of two-dimensional vis-a-vis three-dimensional melting. In two dimensions both the vacancy and the dislocation are point defects, and indeed the former can actually decay by generating a pair of dislocations on close-packed lines separated by one atomic layer. In three dimensions, on the other hand, the vacancy is a point defect, while the dislocation is a line defect. The topological differences between dislocations in systems with these dimensionalities, and the implications for melting theory, have yet to be fully explored.

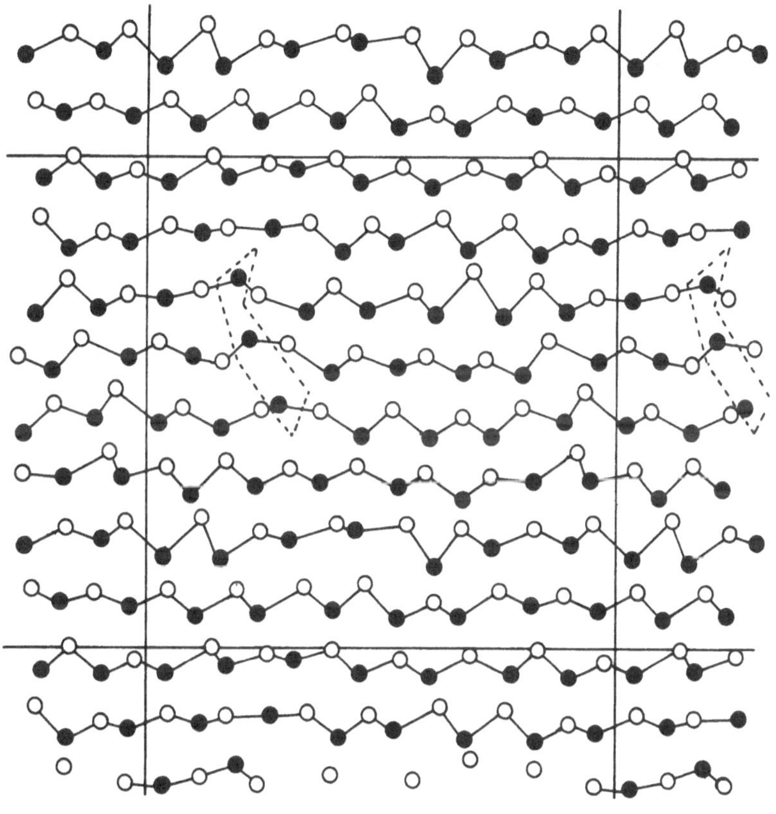

Figure 2. Atomic position in two adjacent close-packed layers, corresponding to State B of Figure 1 (also taken from [14]). Linkage reversal, in which the angle subtended at an atom by its immediate neighbors passes through 180° (changing the zig-zag pattern to zag-zig), permits detection of (Shockley partial) dislocations [12]. The dotted line indicates the approximate position of a dislocation loop found in this manner. The long straight lines indicate the periodic boundaries.

Dislocation-mediated melting belongs to the broader class of melting models based on one form or another of crystal instability. This approach, first suggested by Born [13], is frequently criticized on the grounds that it is essentially a one-phase theory. Evidence has recently been presented that melting does not involve such an instability [14]. A three-dimensional Lennard-Jones model was studied by molecular dynamics, the crystalline state being simulated at various densities along a particular isotherm. No instability was encountered for any density within the thermodynamic interval established by Hansen and Verlet [15] for melting of this system. This result, which is consistent with the view that there is no intrinsic correlation between the properties of the crystalline and liquid phases, is shown in Figure 1. Upon expansion to a density somewhat lower than that of the liquid at freezing, for the same temperature, an instability was in fact encountered. This is apparently a limit of mechanical stability and, as shown in Figure 2, it was found to be dislocation mediated. No dislocations were detected for any density within the thermodynamic melting interval, in conflict with the dislocation theory of melting. By monitoring pair and triplet distribution functions, it was established that the liquid state is produced as a result of the instability. The system could not be made to revert to the crystalline state, upon increasing the density again, and progressive compression within the density interval of thermodynamic melting actually produced a liquid-glass transition. The existence of a virtual glass point in the thermodynamic interval is intriguing. It suggests that melting might occur when the free-energy expenditure in disordering is just offset by the free-energy gain due to increased atomic mobility. Survival of the crystal at a density beyond the freezing volume shows that disorder is a prerequisite for such higher mobility, while the existence of the liquid-glass transition reveals the connection between mobility and density in the disordered state.

3. LOCAL FREE VOLUME AND THE CRYSTAL-LIQUID AND LIQUID-GLASS TRANSITIONS

Both the crystal-liquid [16] and liquid-glass [17] transitions have been analyzed in terms of free volume. In particular, Cohen and Grest [17] have suggested that it is local, rather than global, free volume that is important, and they derived a glass transition temperature using the assumption that this is determined by a percolation limit. The aforementioned crystal instability study [14] also probed the role of local free volume. Two density of states spectra were computed: one for the instantaneous potential energy of all the atoms in the model, and the other for the potential energy that each atom would have if it were moved to the largest adjacent region of free volume. These density of states curves were dubbed the "normal" and "hole" spectra (although it should be remembered that a more accurate term for the latter would be

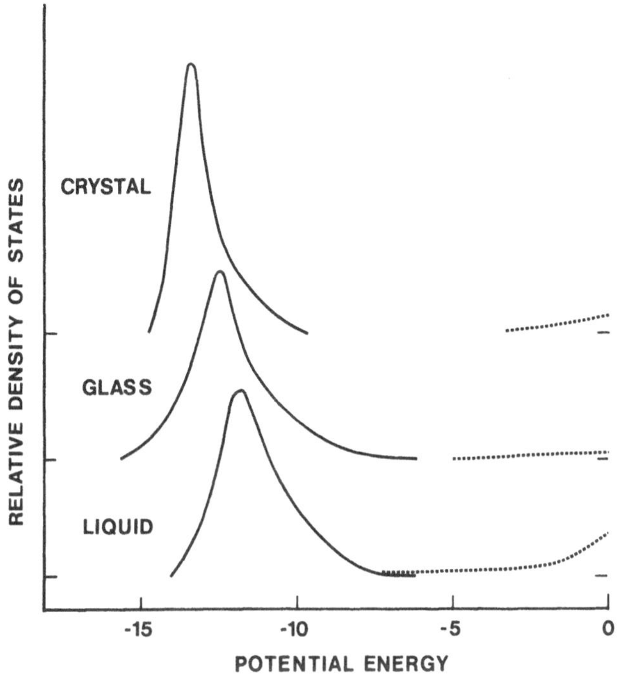

Figure 3. Potential energy density of states spectra for the normal (full curve) and hole (dotted curve) states [14]. The three situations (shifted vertically with respect to one another, for clarity) correspond to the letters A (crystal), F (glass), and E (liquid) indicated in Figure 1, and the curves extend over the ranges for which states were detectable. For the liquid, the maximum of the normal spectrum lies at - 5.97, while the minimum of the hole spectrum occurs at - 7.09. The curves all have the same (arbitrary) ordinate scale.

"atom-in-hole" spectrum). These spectra were computed for the crystalline state at the melting density, the liquid state at the freezing density, and the pressure-induced glassy state in the density interval of thermodynamic melting (see Figure 3). Only for the liquid state were the normal and hole spectra found to overlap, and the distinction between normal and hole states was then no longer well defined. In this case the atoms appear to have sufficient energy to take advantage of the choice of position provided by local free volume, and this is presumably the origin of the enhanced mobility referred to in the previous section.

4. THE BIFURCATION THEORY OF THE STATES OF CONDENSED MATTER

Following an initial suggestion [18] that melting might be the result of a bifurcation instability, it was noted that an interesting experimental result [19] regarding melting entropy might be indicative of bifurcation [20,21]. Stishov, Makarenko, Ivanov, and Nikolaenko [19] investigated the melting entropy, ΔS_m, as a function of the volume change on melting, ΔV_m. Extrapolating to $\Delta V_m = 0$ produced, for a variety of substances, the result $\Delta S_m = R \ln 2$. Because $\ln 1 = 0$, this indicates that whereas a crystal seems to be characterized by the number 1, a liquid appears to be associated with the number 2. This was taken to be suggestive of bifurcation in the liquid state [21], and the R ln 2 was also found to play a role in the liquid-glass, liquid-gas, and even the liquid-superfluid transitions. In short, bifurcation entropy was found to be lost as the liquid undergoes each of these transitions. In the cases of the liquid-gas and liquid-crystal transitions (see Figure 4), and also

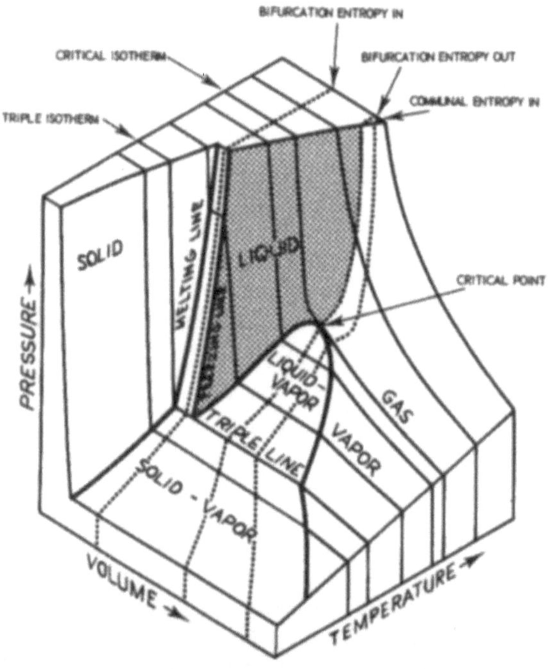

Figure 4. Conjectured modifications to the PVT surface of a simple substance [21]. The distinction between the liquid and gaseous states, the former involving bifurcations while the latter does not, is believed to persist even above the critical temperature.

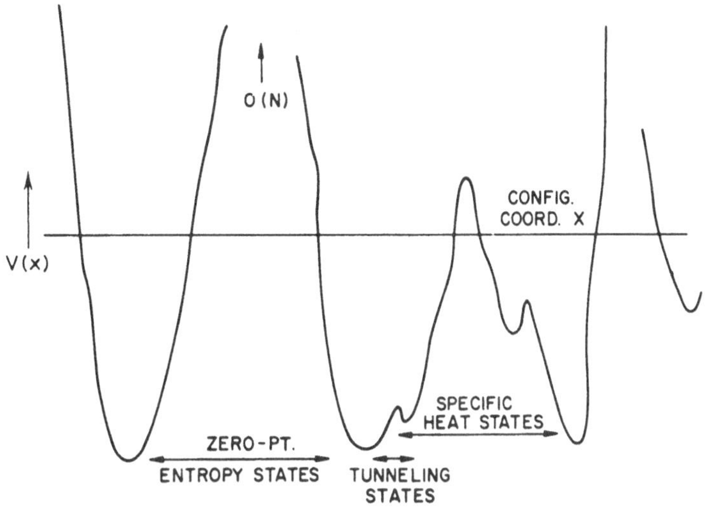

Figure 5. Anderson's "random mountain range" [2]. The tunneling states are believed to be isolated and energetically accessible frozen-in bifurcations [21].

for the liquid-glass transition, this is associated with the disappearance of bifurcation, while for the liquid-superfluid transition it was argued that the bifurcation would still be present, but in a

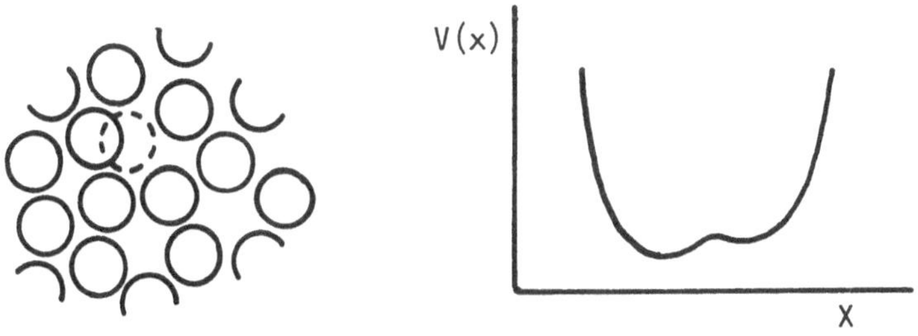

Figure 6. Possible atomic configurations associated with the bifurcation tunneling of an atom in the glassy state, and the corresponding dependence of potential energy on the displacement. Notice that the local symmetry changes for every such jump. The moving atom can jump back, but, alternatively, another atom might move into the space vacated as a result of the first jump. If the latter type of event is sufficiently common, the bifurcations will percolate, and the glass will transform to the liquid state.

special form that has no associated entropy: the so-called Mexican hat state. The two-level systems conjectured by Anderson, Halperin, and Varma [22], to explain the linear dependence of the low-temperature specific heat on temperature, were suggested to be simply frozen-in bifurcations. Thus the lower energy transition indicated in one of the figures appearing in Anderson's review (see Figure 5) were attributed to isolated and energetically accessible bifurcations still present in the low temperature glassy state. One difficulty of the bifurcation theory has been its failure to show why every atom must experience bifurcation, as seems to be required by the entropy R ln 2. In one dimension [18] bifurcation arises from the existence of a point of inflexion in the interatomic potential. In two and three dimensions, the bifurcations would more probably arise from repulsive energy barriers. Figure 6 shows what is likely to be a typical situation in a two-dimensional disordered array of atoms. A favored atom can make a short movement to a close-lying local energy minimum by passing over a low repulsive energy barrier. For a sufficiently expanded crystal the same type of process should be possible, but the disordered state shown in Figure 5 will have the special additional property that the energy minima will lie at different levels; there will be a migrational bias. Moreover, it would appear that the broken symmetry will force the bifurcations to cascade [18], at least if the temperature is sufficiently high, because the moving atom will leave behind a hole into which an adjacent, but different, atom can jump. This superficially resembles the jumping of atoms into and out of a vacancy in a crystal, but in the disordered case there is the vital extra fact that every jump changes the local symmetry and thereby changes the bias on all subsequent jumps. A connection can now be made with the local free volume concept discussed in the preceding section. The migration from an instantaneous site to an adjacent hole is apparently to be identified with a bifurcational jump. If broken symmetry forces the bifurcations to cascade, it might be the case that the "normal" and "hole" spectra, once overlapping, become properties of the system as a whole, and this could be the reason why $\Delta S_m = R \ln 2$, at $\Delta V_m = 0$, and why every atom seems to be effectively bifurcated.

5. THE HARD SPHERE LIMIT

The hard sphere system melts when the volume has expanded to 1.55 times its closest-packed value [23]. It is interesting to note that this number finds a simple explanation in the bifurcation approach. Consider Figure 7, which shows various situations in a hard-sphere face-centered cubic crystal. At maximum density of packing an atom cannot pass out of its coordination shell (see Figure 7b), but if the crystal is sufficiently expanded, with the mean positions of the atoms assumed to lie in their symmetry positions, unhindered escape ultimately becomes possible. By

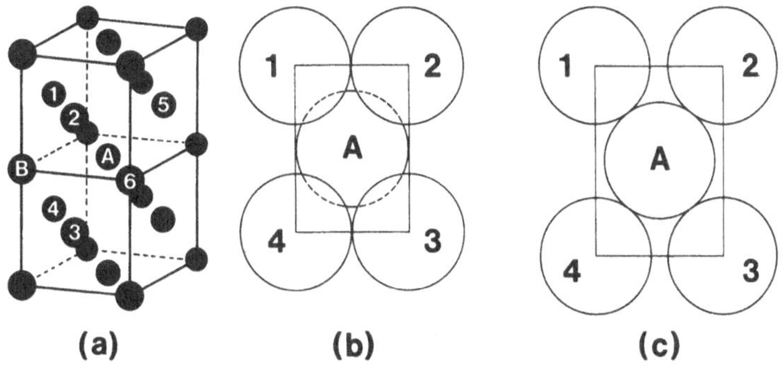

Figure 7. In order to permit an atom, for example Atom A shown in (a), to escape from its coordination shell, the (face-centered cubic) lattice must be expanded. The most open "window" is of the type shown in (b). A volume expansion by a factor of 1.54 is required to just permit the free passage of the moving atom, as shown in (c).

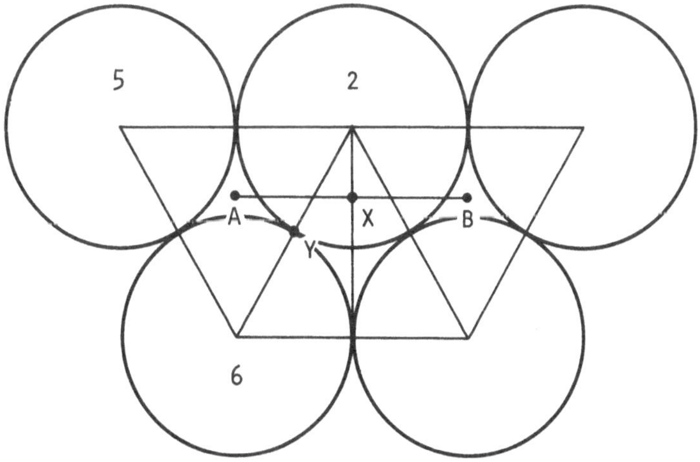

Figure 8. The barriers encountered in a hard sphere model when this is dislocation by the perfect and Shockley partial displacements. A moving atom starts at the position lying above A (in the next layer up), and the barriers are encountered at X and Y, respectively. The numbers and letters are as given in Figure 7(a). A volume expansion by a factor 1.54 is required to permit free passage across the barrier at Y. A larger expansion would be needed for free passage across X.

simple geometry one can demonstrate that the required volume expansion is very close to the factor 1.55 mentioned earlier. The oblong face, which initially has a diagonal of length $3^{1/2}$ d_0, must acquire a diagonal length 2 d_0, d_0 being the sphere diameter. This requires a fractional volume expansion of $\{(2-3^{1/2})/3^{1/2}\}^3$ = 1.54. It so happens that this same expansion is required if two adjacent close-packed planes are to be able to slip with respect to each other. This process, which would produce dislocations of the (Shockley partial) type shown in Figure 2, is indicated in Figure 8. The perfect dislocation in this lattice has a Burgers vector which spans the distance AB (see Figure 8), while the Shockley partial dislocation has a Burgers vector that stretches from A through Y to the equivalent position to A in the next (upward pointing) triangle. Atoms in the next plane above the one shown in the figure lie immediately above A and B, and so on. The perfect Burgers vector displacement would carry the atom A, for instance, up and over position X. The partial Burgers vector displacement would push the same atom over the rather lower barrier at Y, and a straight-forward geometrical calculation shows that the volume expansion required to permit one plane to slip unhindered over the other is precisely the same, namely 1.54, as calculated earlier for hard-sphere bifurcation. Thus the bifurcation and dislocation instabilities actually coincide for the hard sphere limit. By comparison with what was stated in connection with Figures 1, 2, and 3, however, one would anticipate that bifurcations will become operative at a higher density in the disordered state. The thermodynamic interval for melting will probably, in this case too, lie at a higher density than the crystal stability limit.

6. QUANTUM MELTING AND SUPERFLUIDITY

The bifurcation "theory" of quantum melting and superfluidity [21] was based on the experimental observation that the entropy change associated with the liquid-superfluid transition of He^4 is close to R ln 2. Since the superfluid thus lacks the bifurcation entropy, one concludes either that it is not bifurcated or that it is indeed bifurcated but without the additional entropy usually attendant on that condition. It was argued [21] that the latter is the case and that the bifurcated on-site potential has the special form which, in the two-dimensional case, resembles a Mexican hat, the extension to three dimensions being obvious. Unlike for the single-well situation that normally obtains in a crystal, this symmetrically bifurcated state has a $\psi\psi^*$ (where ψ is the wave function) that maximizes elsewhere than the center of gravity: in two dimensions it would be toroidal. Because it is bifurcation that underlies the fluidity of the normal liquid state, this retention of bifurcation, albeit in a special and symmetrical form, by the superfluid is believed [21] to produce its remarkable flow properties. Just how this is

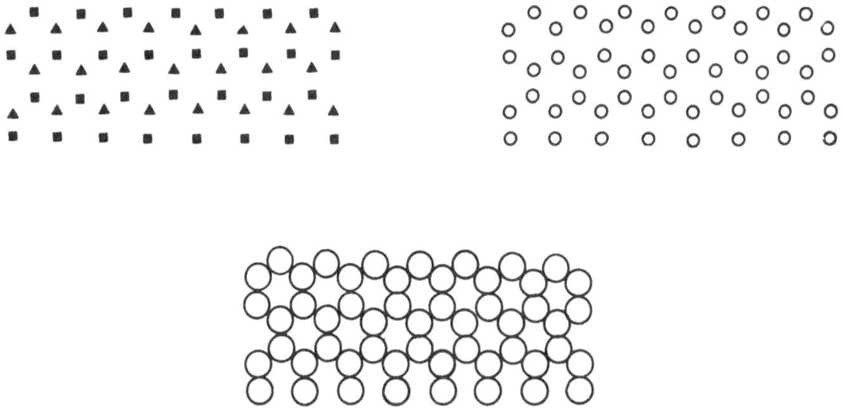

Figure 9. Conjectured situations in a normal face-centered cubic crystal near the absolute zero temperature (upper left), crystalline He^4 (upper right), and superfluid He^4 (lower figure). In the upper left figure the two symbols indicate the atomic positions in two adjacent close-packed layers, and the size of these symbols is without physical significance. In both of the other figures the perimeters of the circles indicate the positions of the minima in the symmetrically bifurcated on-site potential. The maxima of $\psi\psi^*$ will lie near the same radii, and in the case of the superfluid this would give a finite probability that Shockley partial dislocations, of the type shown in Figure 2 (and discussed in connection with Figure 8), are spontaneously generated. (The apparent touching of the circles, in the lower figure, has no significance, because the atoms are on two separate planes.)

accomplished, however, is not yet clear. Moreover, it appears that even crystalline He^4 is unusual in that it too has bifurcated on-site potentials [24]. A possible reconciliation of this fact with the general ideas put forward earlier [21] is illustrated in Figure 9. At the upper left, two adjacent close-packed planes in the face-centered cubic lattice are shown, the time-averaged positions of the atomic nuclei being indicated by two different symbols (the physical size of which is without significance). Thermal vibrations give instantaneous displacements from these positions, of course, but are insufficient to carry atoms over the energy barrier (of the type indicated in Figure 8). In the upper right part of Figure 9, which is conjectured to be the crystalline state of He^4, the symbols have a different significance. The circles indicate the toroidal form of $\psi\psi^*$, the radius corresponding to the maximum of that quantity. In the lower figure, which is believed to represent the superfluid state, the maxima of the toroids now lie at such a large radius that a new situation will prevail: there is a finite probability that the barrier to dislocation generation (as illustrated in Figures 2 and 8) is already bypassed. It is as if the crystal were simultaneously deformed and not deformed, and this might be the reason for the superfluid's remarkable ability to flow. Several pieces of evidence appear to endorse this picture. Both Gross [25] and Overhauser [26] have shown that, for He^4, spatially periodic solutions in the Hartree-Fock approximation have a lower energy than the homogeneous density solution. X-ray scattering [27] and neutron scattering [28] studies both show that the superfluid state has a pair distribution function that closely resembles that of a normal simple liquid. Although the centers of gravity of the atoms in the superfluid model shown in Figure 9 are on a crystal lattice, the maxima of $\psi\psi^*$ are sufficiently removed therefrom that a liquid-like pair distribution function could ensue. Because the centers of gravity are ordered, an additional factor is required to take advantage of the possibility of barrier bypass mentioned earlier. It is possible that the vital agency is the soliton displacement that has already been discussed in connection with the double-well problem [29, 18]. In this connection, it is interesting to note that solitons have been discussed in the context of superfluid He^4 [30].

7. CONCLUSION

Very few numbers have appeared in this short communication, and the discussion has been qualitative. The studies cited in the first three sections, to support the various premises, were based on computer simulation, however, and in this respect the general approach could be claimed to be on a reasonably quantitative footing. There remains the (presumably) more difficult task of developing analytical treatments of the various issues. Until this is achieved, and numerical tests performed, the story must remain little more than speculation.

A particularly noticeable shortcoming of the bifurcation idea, at least in its present form, is that it is essentially a single-particle approach. Many-body effects are implicitly included in that several surrounding atoms will move cooperatively so as to open up both the bifurcation "window" and the local free volume that lies beyond. Nevertheless, the treatment is rather similar to that applied to vacancy jumping in a crystal, and in this respect it might prove unacceptably facile. Weisskopf recently demonstrated that the excitations in liquids involve cooperative motions of a few tens of atoms [31]. The actual nature of these excitations, which remains a mystery, probably holds the key to the liquid problem. It might transpire that it really is just a question of "all for one" (and, by implication, "one for all"), in which case the bifurcation idea will gain credence. If, on the other hand, it is "all for all", with no simplifying quasiparticle approach permissible, the quest for unification of the five states enumerated in the title of this review is going to be an uphill struggle.

Acknowledgments

I am indebted to my colleagues J.U. Madsen and J.L. Tallon, joint work with whom clearly forms the backbone of the present article. Thanks are also due to L. Chapoy, M. Doyama, S.F. Edwards, J. Finney, A. Luther, T. Ninomiya, N.L. Peterson, and V.F. Weisskopf for stimulating discussions, and to the Danish Natural Science Research Council (Statens Naturvidenskabelige Forskningsråd) for financial support.

REFERENCES

[1] M. Lazzari, B. Scrosati, and C.A. Vincent, J. Amer. Ceramic Soc. 61, 451 (1978).

[2] P.W. Anderson, in Proc. Les Houches Conf. on Ill-Condensed Matter, eds. R. Balian et al. (North Holland, Amsterdam, 1979) p.162.

[3] N.F. Mott and R.W. Gurney, Trans. Faraday Soc. 35, 364 (1939).

[4] W.L. Bragg, in Proc. Symp. on Internal Stresses (Institute of Metals, London, 1947) p.221.

[5] W. Shockley, in l'Etat Solide (Inst. Inter. de Physique Solvay, Brussels, 1952) p.431.

[6] A. Ookawa, J. Phys. Soc. Japan 15, 2191 (1960).

[7] S. Mizushima, J. Phys. Soc. Japan 15, 70 (1960).

[8] D. Kuhlmann-Wilsdorf, Phys. Rev. A 140, 1599 (1965).

[9] R.M.J. Cotterill, in Ordering in Strongly Fluctuating Condensed Matter Systems, ed. Tormod Riste (Plenum Publishing Corporation, 1980) p.261.

[10] R.M.J. Cotterill and L.B. Pedersen, Solid State Comm. 10, 439 (1972).

[11] J.M. Kosterlitz and D.J. Thouless, J. Phys. C 6, 1181 (1973).

[12] R.M.J. Cotterill, W. Damgaard Kristensen, and E.J. Jensen, Phil. Mag. 30, 245 (1974).

[13] M. Born, J. Chem. Phys. 7, 591 (1939).

[14] R.M.J. Cotterill and J.U. Madsen, Nature (in press).

[15] J.P. Hansen and L. Verlet, Phys. Rev. 184, 151 (1969).

[16] D. Turnbull, in Liquids: Structure, Properties, Solid Interactions, ed. T.J. Hughel (Elsevier, Amsterdam, 1965) p. 6.

[17] M.H. Cohen and G.S. Grest, Phys. Rev. B 20, 1077 (1979).

[18] R.M.J. Cotterill, Physica Scripta 18, 37 (1978).

[19] S.M. Stishov, I.N. Makarenko, V.A. Ivanov, and A.M. Nikolaenko, Phys. Lett. 45A, 18 (1973).

[20] R.M.J. Cotterill, J. Cryst. Growth 48, 582 (1980).

[21] R.M.J. Cotterill and J.L. Tallon, Discussions of the Faraday Society No. 69 (in press).

[22] P.W. Anderson, B.I. Halperin, and C.M. Varma, Phil. Mag. 25, 1 (1972).

[23] T.E. Faber, An Introduction to the Theory of Liquid Metals (Cambridge University Press, 1972) p. 94.

[24] R.A. Guyer, Solid State Physics 23, 413 (1969).

[25] E.P. Gross, Phys. Rev. Lett. 4, 57 (1960).

[26] A.W. Overhauser, Phys. Rev. Lett. 4, 415 (1960).

[27] W. Gordon, C. Shaw, and J. Daunt, J. Phys. Chem. Solids 5, 117 (1958).

[28] D.G. Henshaw, Phys. Rev. 119, 9 (1960).

[29] J.A. Krumhansl and J.R. Schrieffer, Phys. Rev. B 11, 3535 (1975).

[30] S. Nakajima, S. Kurihara, and K. Tohdoh, J. Low Temp. Phys. 39, 465 (1980).

[31] V.F. Weisskopf, Trans. New York Acad. Sci. 38, 202 (1977).

MOLECULAR DYNAMICS STUDIES OF SUPERIONIC CONDUCTORS*

A. Rahman and P. Vashishta

Argonne National Laboratory
Argonne, Illinois 60439

ABSTRACT

Structural and dynamical properties of superionic conductors AgI and CuI are studied using molecular dynamics (MD) techniques. Our model of these superionic conductors is based on the use of effective pair potentials. To determine the constants in these potentials, cohesive energy and bulk modulus are used as input; in addition one uses notions of ionic size based on the known crystal structure.

Salient features of the MD technique are outlined. Methods of treating long range Coulomb forces are discussed in detail. This includes the manner of doing Ewald sum for MD cells of arbitrary shape. Features which can be incorporated to expedite the MD calculations are also discussed.

A novel MD technique which allows for a dynamically controlled variation of the shape and size of the MD cell is described briefly. The development of this novel technique has made it possible to study structural phase transitions in superionic conductors. For α-AgI, among the structural properties we have studied are: partial pair correlation functions, mean square displacements of iodines, cation density maps, Havens ratio, etc. The dynamical properties examined include cation self-diffusion, nature of cation jumps, bias in successive jumps, velocity auto correlation functions, current-current correlation functions.

---

*Work supported by the U.S. Department of Energy.

In CuI, we have examined the microscopic nature of $\gamma \to \alpha$ transition. It is found that at about 700 K the copper ions undergo an order-disorder transformation leading to a specific heat anomaly. The nature of the first-order transition and its precursor effects are also analyzed. Results for a number of other structural and dynamical properties for α-CuI are presented.

In AgI the $\alpha \rightleftarrows \beta$ transition is studied using the new MD technique. In our model, upon heating β-AgI, the iodines undergo hcp$\to$bcc transformation and the silver ions become mobile, whereas the reverse transformation is observed on cooling α-AgI.

INTRODUCTION

Over the last 25 years molecular dynamics (MD) and Monte Carlo (MC) calculations (computer simulations) have become widely used techniques for the study of condensed systems. Even though the largest amount of effort has gone into the study of fluids, considerable work on perfect solids at high temperature and on solids with defects has also been done. It is in the study of superionic conductors that features of both fluid and solid (MD and MC) calculations come together in a most interesting manner.

AgI and CuI are among the most widely studied superionic conductors. The electrical conductivity of α-AgI was experimentally determined by Tubandt and Lorenz[1] as early as 1914. More recent experimental observations on AgI include infrared spectra, Raman and Brillouin spectra, microwave absorption, X-ray and neutron diffraction studies, "Extended X-ray Absorption Fine Structure" (EXAFS) experiments and inelastic neutron scattering studies etc.[2-8] In constructing realistic models to explain many of the above mentioned experimental observations several difficulties are encountered. Major difficulty lies in including large anharmonic vibrations of iodine ions away from their lattice sites. Even more difficult is the inclusion of the effects due to collective motions of the cations themselves.

A motivation for the work presented here is to investigate if it is possible to describe structural and dynamical properties of superionic conductors such as AgI and CuI using effective pair wise potentials.[9-14] We have studied the nature of ionic motions in AgI and CuI using molecular dynamics technique.[13,14] Effective pair potential functions are constructed using the simple concept of ionic size.[15] The constants in the potential functions can be determined using the bulk properties of the system, e.g., crystal structure, cohesive energy and bulk modulus.[16,17] A variety of structural and dynamical properties are calculated for AgI and CuI. Among structural properties we have studied partial pair correlation functions, mean square displacement of iodines, cation density maps, Havens ratio, etc. The dynamical properties that we have examined include cation self diffusion, nature of jumps, successive cation jumps, velocity auto correlation functions, current-current correlation functions, nature of collective modes etc.[13,14,18,19]

In CuI we have also studied the nature of $\gamma \rightarrow \alpha$ transition.[14] It is found that at about 700 K the copper ions undergo an order-disorder transformation leading to a specific heat anomaly. The nature of the first-order transition and its precursor effects are also analyzed.

Using a novel molecular dynamics technique introduced by Parrinello and Rahman[20] we have very recently studied the structural phase transitions in these superionic conductors. It should be remarked here that a study of the structural transitions (e.g. structural changes of the iodine lattice) is impossible within the framework of the old MD and MC technique. In particular, we have studied the $\alpha \rightleftarrows \beta$ phase transitions in AgI using the new MD technique which allows changes in the shape and size of the periodically repeating cell.[21] In our model, upon heating β-AgI, the iodines undergo hcp$\rightarrow$bcc transformation and the silver ions become mobile, whereas the reverse transformation is observed on cooling α-AgI.

In this presentation we shall first mention the overall features of computer simulations in a very brief manner. The problems related with long range Coulomb interactions will be discussed next. More recently approximate methods have been introduced to reduce the computational expense related with the use of Ewald summations in dealing with r^{-1} potentials. Apart from the Coulomb potential, the choice of the short range repulsive and attractive potentials is always a difficult problem; points of view related with the solution (or circumvention!) of this problem will be discussed.

The method of molecular dynamics was outlined in lectures given at another NATO Study Institute[22] and it will be assumed that the reader will have access to those lectures. In those lectures the main focus was on short range interactions and hence Coulomb potentials were not considered at all. In the following we shall discuss the salient features of the molecular dynamics technique; special features which arise when Coulomb forces are present are discussed in detail following this general discussion.

A BRIEF OUTLINE OF THE MOLECULAR DYNAMICS (MD) TECHNIQUE

In MD the purpose is to calculate with a desired degree of precision the classical trajectories of a system of N particles interacting with one another in some precisely defined manner. If the coordinates of the particles are denoted by $\underline{r}_i$, their masses by m_i, i=1,...,N and if V ($\underline{r}_1$,...,$\underline{r}_N$) is the potential energy of the system, Newton's equations are

$$m_i \ddot{\underline{r}}_i = - \underline{\nabla}_i V, \; i = 1,2,...N.$$

To solve these equations one needs to specify initial positions and velocities of all the particles. The integration of the equations then gives the trajectories of all the particles for subsequent (and earlier!) times. When the integration is performed by using a suitably chosen algorithm for the numerical integration of coupled, non-linear, ordinary, second order differential equations the ensuing dynamics should lead to a conservation of the total energy E $(=V+1/2 \; \Sigma \; m_i \; \dot{\underline{r}}_i^2)$ of the system and this can be used as a criterion for judging the accuracy of the trajectories generated.

A few remarks on terminology are relevant at this point. Since one is concerned with the statistical mechanics of classical N- body systems it is usual to rewrite the kinetic energy of the system $1/2 \; \Sigma \; m_i \; \dot{\underline{r}}_i^2$ as $3/2 \; N \; k_B T$. Conservation of total energy requires $V + 3/2 \; N \; k_B T$ to be constant, while V and T change with time. Moreover, if on applying suitable tests the system is considered to be in statistical equilibrium, the time average of T (t) over a long trajectory of the system is usually referred to as the temperature of the system and in the same spirit T (t) is referred to as the "temperature" at time t.

In a state of equilibrium there will emerge from such calculations a precise relation T (E) [or equivalently E (T)] between the total energy of the system and its temperature (averaged over a long enough trajectory), and this relation defines the equation of state of that system. Again using the language of statistical mechanics, one can go from one energy shell to another by suitable manipulation at any moment of the $\dot{\underline{r}}_i$'s i.e. of T (t). A system can be heated (or cooled) by applying a factor $\alpha > 1$ (or <1) to all velocities at time t. Obviously after any such manipulation a certain time has to be allowed to pass before a new equilibrium state is established. It is of course not excluded that the purpose of moving the system from one energy shell to another is precisely to study the details of the manner in which the new equilibrium is achieved.

CHOICE OF THE ALGORITHM

The problem is to convert the differential equations describing the motion of the particles into a set of difference equations which enable us to go from time t to $t+\Delta t$ with a suitably chosen Δt. We shall consider later the value to be chosen for Δt.

In the literature that has grown over the last fifteen years or so, the simplest algorithm is that used by Verlet[23] and his collaborators at Orsay, France. In this algorithm the positions of the particles are written as a Taylor series expansion:

$$\underline{r}_i(t\pm\Delta t) = \underline{r}_i(t) \pm \Delta t\, \dot{\underline{r}}_i(t) + 1/2\, (\Delta t)^2\, \ddot{\underline{r}}_i(t) \pm \cdots$$

giving the expression:

$$\underline{r}_i\,(t+\Delta t) = 2\underline{r}_i(t) - \underline{r}_i(t-\Delta t) + (\Delta t)^2\, \ddot{\underline{r}}_i(t) + 0[(\Delta t)^4]$$

If m_i is the mass, the acceleration, $\ddot{r}_i(t)$, of the coordinate $\underline{r}_i(t)$ is given by Newton's equation of motion.

Thus, $\ddot{\underline{r}}_i(t)$ can be calculated from the knowledge of all the coordinates at time t i.e., of all the $\underline{r}_i(t)$. Thus the only information that needs to be carried in memory is $\underline{r}_i(t-\Delta t)$ and $\underline{r}_i(t)$ for all i. Using all $\underline{r}_i(t)$ we first calculate all the $\ddot{\underline{r}}_i(t)$ from the equation of motion and having done that all the $\underline{r}_i(t+\Delta t)$ from Verlet's algorithm.

To get the velocities at time t, so as to calculate the kinetic energy at that time, the simplest procedure is to calculate

$$[\underline{r}_i(t+\Delta t) - \underline{r}_i(t-\Delta t)]/2\Delta t\ ,$$

as a measure of the velocity at time t.

Whatever the algorithm used, the process of calculating $r_i(t)$ can obviously be combined with that of calculating V itself so that at the end of the process of going from t to $t+\Delta t$ we have the kinetic and the potential energy at time t.

A predictor-corrector type algorithm which has been used by various authors can be found in a report by Gear.[24] This algorithm makes use of the derivatives of $\underline{r}_i$ up to a predetermined order, say 5, for example, and all these derivatives need to be carried in the memory. Let us denote any given cartesian component of $\underline{r}_i(t)$ by q_o, of $\dot{\underline{r}}_i(t)\Delta t$ by q_1, of $\ddot{\underline{r}}_i(t)(\Delta t)^2/2$ by q_2, and so on. Using a Taylor expansion it is possible to predict the values of all the derivatives of the positions at $t+\Delta t$. Denoting the predicted values of q_i by p_i one obtains

$$
\begin{aligned}
p_o &= q_o + q_1 + q_2 + q_3 + q_4 + q_5 \\
p_1 &= q_1 + 2q_2 + 3q_3 + 4q_4 + 5q_5 \\
p_2 &= q_2 + 3q_3 + 6q_4 + 10q_5 \\
p_3 &= q_3 + 4q_4 + 10q_5 \\
p_4 &= q_4 + 5q_5 \\
p_5 &= q_5
\end{aligned}
$$

Hence from p_2 we can get the predicted value of the acceleration at $t+\Delta t$.

Using p_o, which are the predicted positions at $t+\Delta t$ the accelerations in predicted positions can be calculated using the equations of motion; let $a(p_o)$ denote these accelerations and $\tilde{p}_2$ denote $a(p_o)(\Delta t)^2/2!$. The difference $D = \tilde{p}_2 - p_2$ allows us to get corrected values c_i from the predicted values p_i as follows:

$$c_i = p_i + f_{i2}^{(5)} D \ , \ i = 0,1,\ldots,5$$

where the magic numbers $f_{i2}^{(5)}$ are 3/16, 251/360, 1, 11/18, 1/6, 1/60, respectively for $i = 0,1,\ldots,5$. In $f_{ij}^{(k)}$ j denotes the order of differential equations being solved, (k) the order of the highest derivative being used in the algorithm and i goes from 0 to k. The reason for choosing three indices to denote these magic numbers is that the values of $f_{ij}^{(k)}$ depend on all three indices.

It is important to note that for clarity of presentation we have used the symbols p and c. In fact, the only memory locations needed are for the q's and for the accelerations $a(p_o)$ mentioned above.

The predictor-corrector method has the advantage that one can use the differences c_0-p_0 and c_1-p_1 to put bounds on the acceptable error. The predictor-corrector loop can be repeated if necessary by using the corrected values as the predicted values and going back to the calculation of $a(c_0)$ and then D would be $\tilde{c}_2-c_2, \tilde{c}_2$ being $a(c_0)x(\Delta t)^2/2!$.

This algorithm also allows solutions of a coupled set of first and second order differential equations (e.g. the Newton-Euler equations of motion for rigid bodies) in the same manner as explained above.[25] As for the problem of initiating the calculation, a practical scheme would be to put all the derivatives from 2 upwards equal to zero; in a few steps of Δt the algorithm itself generates appropriate values of these derivatives. More sophisticated methods of starting a calculation can also be used but in the context of statistical mechanics the "error" committed in the first few Δt has no significance for the final results which are normally averages taken from runs stretching over thousands of time steps Δt of integration.

CALCULATION OF THE FORCES

This being the most time consuming part of the whole calculation, particular attention has to be paid so as to make it as efficient as possible. Let us assume that the potential V is given to be a sum of 2-body potentials so that one can write

$$V = 1/2 \sum_{ij} \phi(r_{ij})$$

The force on particle i due to particle j is

$$-\underline{\nabla}_i \, \phi_{ij} = -\frac{1}{r_{ij}} \frac{d\phi_{ij}}{dr_{ij}} \, \underline{r}_{ij}$$

where $\underline{r}_{ij} = \underline{r}_i - \underline{r}_j$. Thus, we need, in analytic or tabulated form, the two functions ϕ and ϕ'_{ij}/r_{ij} for all pairs ij. Note that we can drop the indicies ij on ϕ_{ij} when all particles are identical; otherwise as many functions are needed as there are distinct pairs. Also, if ϕ depends only on even powers of r then the function needed for the calculation of the forces also has only even powers which makes for considerable saving since no square root calculation needs to be performed. However for any function ϕ_{ij} one can use this advantage (of avoiding the square root calculation) by tabulating the function not as a function of r but of r^2. Eventually each such advantage has to be balanced between memory requirements for finely meshed tables to avoid involved interpolation arithmetic against the cost of interpolating in a not so finely meshed tabulation.

The calculation of the forces thus proceeds by programming the double sum with $j > i$ and without 1/2, calculating $\underline{r}_{ij}$, its square r_{ij}^2, the value of ϕ_{ij} and $r_{ij}^{-1}\,\phi_{ij}'$ and finally the three components of the force on i due to j. The same numbers with changed sign give the force on j due to i.

While r_{ij}^2 is available to make these calculations, it is obviously useful to evaluate other properties of the system which depend on the distance. The most important is the pair correlation function. Note that if the system is composed of k different kinds of particles there will be $k(k-1)/2$ different pair potentials and pair correlations.

TRUNCATION OF THE POTENTIAL AND LISTS OF INTERACTING NEIGHBORS

In many cases it is worthwhile to work with a truncated potential. Symbolically we write

$$\phi_{ij}(r) = \phi_{ij}(r_c) \quad \text{for} \quad r \geq r_c \ ;$$

the most famous example of this is the truncated Lennard-Jones potential made famous by the perturbation theory of liquids.[26] In that case one takes r_c to be the value ("bottom of the well") at which $\phi'(r) = 0$; this makes the truncation smooth in the derivative of ϕ as well. However, even when the attractive part of the potential is to be included the potential may be truncated at a suitably large value.

In either case, the following method invented by Verlet[23] is useful when dealing with moderately large systems. For very small systems ($n = 100$ to 200) such procedures are not of any use, and for very large systems ($n = 5000$ up) further elaboration of the procedure becomes necessary.[27]

Unless one is dealing with very dilute systems (for which MD is a doubtful method of investigation anyway) one can state that for several steps Δt of integration, the neighbors up to r_c will be in large majority, unchanged. A few move out of range (i.e., r_c) and a few will move within range. Thus, if at any moment we construct a list of neighbors not upto r_c but upto $r_c + S$, where S denotes a "skin" thickness, then for several Δt after that moment, we need to consult only this list (and not the whole system) to identify neighbors that are within range r_c of interaction and to throw away those beyond r_c.

A little consideration will show that there is an optimum balance between the value of S and the number of Δt for which the list, once made , may be used. We shall not dwell on this. However, we shall mention the fact that the list array should be a one dimensional array with an auxiliary array of dimension N; this latter array is a marker array which provides indicators for the part of the list relevant for each of the N particles. The following diagram will make this obvious.

# of element of list	1	2	3	4	5	.	.	.	L
content of register	a	b	c	d	e	.	.	.	.
# of element of markers array	1	2	3	4	5	.	.	.	.
content of register	k	m	n	o	p	.	.	.	.

L is the number of locations reserved for the list and N is the number of particles. a,b,c etc. are the identities of particles, i.e. they are any of the tags 1 to N and are modified every time the list is remade. k=0 always, m is that element of the total list where the sublist of neighbors interacting with particle #1 ends and at m+1 the sublist of neighbors interacting with #2 starts.

PERIODIC BOUNDARY CONDITIONS

We shall consider the usual periodic boundary conditions (pbc) which enable us to extend a parallelopiped box to infinity by an integral number of translations in the three directions which determine the shape and size of the box. It is customary, while working on liquids, to use a cubic box; however, when dealing with the dynamics of crystal of non-cubic symmetry it is more convenient to work in a non-orthogonal system and to calculate the squares of distances, for example, with appropriate cross terms. However, we shall not consider this further assuming that the N particles are confined in a cubic box of length L repeated indefinitely by translations in the three orthogonal directions.

The pbc have the merit of removing the surface effects in a mathematically well defined manner. Intuitively it seems correct that in the case of potential functions which are short ranged compared to half the box size the effect of pbc will be rather small. For long range potentials like the Coulomb interaction this may not be so and it appears that systematic analysis of such effects has not yet been made. The work done so far on the one component plasma and on molten salts using 200 and 1000 particles has given no indication of systematic effects arising out of pbc. Mandell[28] was able to conclude from his molecular dynamics work on small systems using the (6-12) potential that pbc induce certain 3-body correlations and systematic effects can also be seen in the value of the virial.

In visualizing a system with pbc it may be convenient to think in terms of a "box" with walls, and particles "entering" from one face while "leaving" from the opposite face. However, since the box can be drawn anywhere in space, as long as it has the correct size and is never tilted, one can always think of any particle as being at the center of the box rather than at one face or another. Both ways of visualizing the pbc are of course equally valid. When dealing with a short range potential and a range of interaction r_c ($<L/2$) the two pictures will appear in a hand-drawn sketch as:

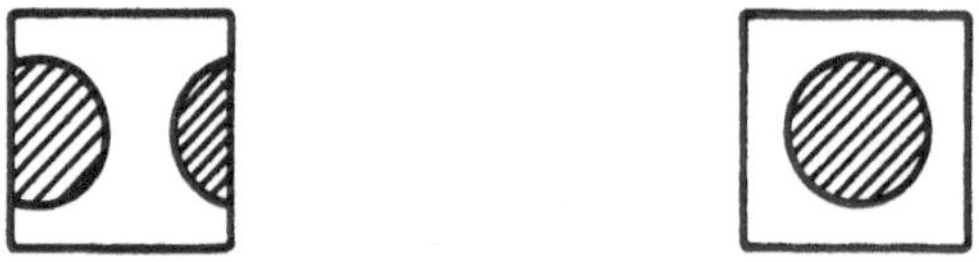

The shaded region is a sphere of radius r_c. In a computer program with coordinates given with respect to a fixed origin (usually taken at the center of the box) it is the picture on the left which will be in operation. While checking to see if two particles i, j are within range each component x_{ij}, y_{ij}, z_{ij} of $\underline{r}_{ij} = \underline{r}_i - \underline{r}_j$ has to be tested separately; if $|x_{ij}| < r_c$ one proceeds to y_{ij} but otherwise one takes $x_{ij} = x_{ij} - L[x_{ij}/|x_{ij}|]$ as the x coordinate difference and the test is made again. This is equivalent to slicing a plate (of thickness $2 \times r_c$), perpendicular to the x axis before looking at y and z. After completing this operation for y the sum of squares is tested against r_c^2 and the z coordinate is tested only if this sum is $< r_c^2$. Finally the pair is accepted for interaction if the distance squared is $< r_c^2$. It should be noted that the sign $[x_{ij}/|x_{ij}|]$ can be determined without arithmetic by simply remembering whether i is in the left or the right half of the box. Also, for L much larger than $2r_c$ (say $3r_c$ or more) the slicing can be made more efficient by including, after the test $|x_{ij}| < r_c$, the additional test $|x_{ij}| > L-2r_c$ before applying the translation and testing x_{ij} against r_c.

When a particle coordinate goes beyond $\pm L/2$ it has to be reset so as to bring it back into the box from the opposite face. This operation simply recognizes that the image is already in the box and transfers the particle tag to its image.

ENERGY AND FORCES DUE TO COULOMB INTERACTIONS

We shall consider the case of a cubic box of volume $\Omega = L^3$ which contains N ions. These carry charges q_i and are located at $\underline{r}_i$, $i = 1,\ldots,N$. We assume $\Sigma q_i = 0$. The box is repeated periodically so that an infinite system is generated with particles of charge q_i at $\underline{r}_i + L(\lambda, \mu, \nu)$, the integers λ, μ, ν going from minus to plus infinity.

The electrical potential of particle at $\underline{r}_i$ is then

$$V_i = \sum_{j=1}^{N} \sum_{\underline{n}}' \frac{q_j}{|\underline{r}_{ij}+L\underline{n}|}$$

where $\underline{n}$ indicates the triplet of integers (λ, μ, ν) and Σ' indicates the absence of the $\underline{n} = 0$ term when $j = i$. The electric field at the position of this particle is then obtained by differentiating with respect to $\underline{r}_i$.

The problem of the convergence of this series has generated considerable literature and has recently been investigated anew by de Leeuw, Perram and Smith[29] who have shown that the problem of summing the series to infinity cannot be decoupled from the boundary conditions. The results that we will write down below are a special case of the results of the above authors and arise from their expressions when, so to say, "the sample is wrapped in tinfoil!"

Going back to the expression for V_i above, the total potential energy of the N charged particles will then be

$$V_N = + \frac{1}{2} \sum_{i=1}^{N} q_i \sum_{j=1}^{N} q_j \sum_{\underline{n}}' \frac{1}{|\underline{r}_{ij}+L\underline{n}|} .$$

To calculate V_N we may proceed as follows: in the unit cell of volume Ω the charge density is

$$n(\underline{r}) = \sum_{j=1}^{N} q_j \, \delta(\underline{r}-\underline{r}_j)$$

with the condition $\Sigma q_i = 0$. We define $n(\underline{k})$ by

$$n(\underline{k}) = \int_{\Omega} n(\underline{r}) \, e^{-i\underline{k}\cdot\underline{r}} \, d\underline{r} = \sum q_j \, e^{-i\underline{k}\cdot\underline{r}_j}$$

Inverting the definition of $n(\underline{k})$ we get

$$n(\underline{r}) = \frac{1}{\Omega} \sum_{\underline{k}} n(\underline{k}) \, e^{i\underline{k}\cdot\underline{r}} = \frac{1}{\Omega} \sum_{\underline{k}} \sum_{j} q_j \, e^{i\underline{k}\cdot(\underline{r}-\underline{r}_j)}$$

Periodicity of the distribution implies that in the summation over $\underline{k}$ only vectors of the mesh $2\pi/L$ (λ, μ, ν) are present (in other words the reciprocal lattice vectors have to satisfy the Bragg condition for a simple cubic lattice of lattice constant L). For any other $\underline{k}$, $n(\underline{k}) = 0$.

From the fact that at any point in space the electric potential ϕ satisfies $\nabla^2\phi(r) = -4\pi n(r)$, we can obtain $\phi(k)$ the Fourier component of $\phi(r)$.

$$\text{Let } \phi(r) = \frac{1}{\Omega}\sum_{\underline{k}} \phi(\underline{k})\, e^{i\underline{k}\cdot\underline{r}} .$$

Substituting in the Poisson equation,

$$-\frac{1}{\Omega}\sum_{\underline{k}} \underline{k}^2 \phi(\underline{k})\, e^{i\underline{k}\cdot\underline{r}} = -\frac{4\pi}{\Omega}\sum_{\underline{k}}\sum_{j} q_j\, e^{i\underline{k}\cdot(\underline{r}-\underline{r}_j)}$$

or

$$\phi(\underline{k}) = \frac{4\pi}{k^2}\sum_{j} q_j\, e^{-i\underline{k}\cdot\underline{r}_j} \quad (\underline{k}\neq 0)$$

The potential energy of the charges q_j, $j=1,\ldots,N$, in the cell of volume Ω is then

$$V_N = \frac{1}{2}\int_{\Omega} n(\underline{r})\phi(\underline{r})d\underline{r} - \frac{1}{2}\sum_{j} q_j^2 \int \frac{\delta(\underline{r}-\underline{r}_j)}{|\underline{r}-\underline{r}_j|}\, d\underline{r}$$

$$= \frac{1}{2}\frac{4\pi}{\Omega}\sum_{\underline{k}}{}' \; 1/\underline{k}^2 \sum_{j} q_j\, e^{-i\underline{k}\cdot\underline{r}_j} \int_{\Omega} n(r)\, e^{+i\underline{k}\cdot\underline{r}}\, d\underline{r}$$

$$-\frac{1}{2}\sum_{j} q_j^2 \int \frac{\delta(\underline{r})}{|\underline{r}|}\, d\underline{r}$$

$$= \frac{1}{2}\frac{4\pi}{\Omega}\sum_{\underline{k}}{}' \; 1/\underline{k}^2 \sum_{j} q_j \sum_{\ell} q_\ell\, e^{i\underline{k}\cdot(\underline{r}_\ell-\underline{r}_j)} - \text{the self term.}$$

$$= \frac{1}{2}\frac{4\pi}{\Omega}\sum_{j}\sum_{\ell\neq j} q_j\, q_\ell \sum_{\underline{k}}{}' \frac{e^{i\underline{k}\cdot\underline{r}_{\ell j}}}{k^2} + \frac{1}{2}\sum_{j} q_j^2 \left(\frac{4\pi}{\Omega}\sum_{\underline{k}}{}' \; 1/\underline{k}^2\right.$$

$$\left. - \int \frac{\delta(\underline{r})}{|\underline{r}|}\, d\underline{r}\right).$$

$\sum_{\underline{k}}'$ means that $\underline{k} = 0$ is excluded; writing $\psi(\underline{r}) = 4\pi/\Omega \sum_{\underline{k}}' \dfrac{e^{i\underline{k}\cdot\underline{r}}}{k^2}$ and

using the neutrality condition $\sum_j \sum_{\ell \neq j} q_j q_\ell + \sum_j q_j^2 = (\sum_j q_j)^2 = 0,$

$$V_N = \frac{1}{2} \sum_j \sum_{\ell \neq j} q_j q_\ell \left\{ \psi(\underline{r}_{\ell j}) - \underset{r \to 0}{\text{Lt}} [\psi(r) - \frac{1}{r}] \right\}$$

The expression in { } then can be thought of as an effective Coulomb interaction between ℓ and j situated at $\underline{r}_\ell$ and $\underline{r}_j$ in the cell containing N charges. Note that it depends not on $|\underline{r}_{\ell j}|$ but on $\underline{r}_{\ell j}$.

The Ewald method for evaluating $\psi(\underline{r})$ is as follows: We rewrite $\psi(\underline{r})$ with a convergence factor $e^{-\gamma^2 k^2}$ in $\underline{k}$ space.

$$\psi(\underline{r}) = \frac{4\pi}{\Omega} \sum_{\underline{k}}' \frac{e^{-\gamma^2 k^2} e^{i\underline{k}\cdot\underline{r}}}{k^2} + \frac{4\pi}{\Omega} \sum_{\underline{k}}' \frac{(1-e^{-\gamma^2 k^2})\, e^{i\underline{k}\cdot\underline{r}}}{k^2}$$

But $(1-e^{-\gamma^2 k^2})/\underline{k}^2 = \dfrac{1}{8\pi^3} \displaystyle\int \frac{2\pi^2}{r} [1-\text{erf}\,(|\underline{r}|/2\gamma)]\, e^{-i\underline{k}\cdot\underline{r}}\, d\underline{r}$. This equality is the key to the Ewald method of summation.

The second term on the right in the expression for $\psi(\underline{r})$ is then

$$\frac{4\pi}{\Omega} \sum_{\underline{k}}' \frac{1}{8\pi^3} \int \frac{2\pi^2}{s} \,\text{cerf}\,(s/2\gamma)\, e^{i\underline{k}\cdot(\underline{r}-\underline{s})}\, d\underline{s}$$

or

$$\int \frac{d\underline{s}}{s} \,\text{cerf}\,(s/2\gamma) \frac{1}{\Omega} \sum_{\underline{k}}' e^{i\underline{k}\cdot(\underline{r}-\underline{s})}$$

or

$$\int \frac{d\underline{s}}{s} \,\text{cerf}\,(s/2\gamma) \left\{ \sum_{\underline{\nu}} \delta(\underline{r}-\underline{s}-\underline{\nu}) - \frac{1}{\Omega} \right\}$$

We have used $\underline{\nu}$ to denote the vector $L\underline{n}$. Thus

$$\psi(r) = \frac{4\pi}{\Omega}\sum_{\underline{k}}{}' \frac{e^{-\gamma^2 k^2}}{k^2} e^{i\underline{k}\cdot\underline{r}} + \sum_{\underline{\nu}} \frac{\mathrm{cerf}(|\underline{r}-\underline{\nu}|/2\gamma)}{|\underline{r}-\underline{\nu}|} - \frac{4\pi\gamma^2}{\Omega}$$

Moreover

$$\operatorname*{Lt}_{r\to 0} [\psi(\underline{r}) - 1/r] = \frac{4\pi}{\Omega}\sum_{\underline{k}}{}' \frac{e^{-\gamma^2 k^2}}{k^2} + \sum_{\underline{\nu}}{}' \frac{\mathrm{cerf}(|\underline{\nu}|/2\gamma)}{|\underline{\nu}|} - \frac{4\pi\gamma^2}{\Omega}$$

$$+ \mathrm{Lt}\left[\frac{\mathrm{cerf}(r/2\gamma)}{r} - \frac{1}{r}\right]$$

and the last term on the right is $-1/\gamma\sqrt{\pi}$.

Hence the effective Coulomb interaction, apart from the $q_j q_\ell$ in front becomes

$$\sum_{\underline{\nu}} \frac{\mathrm{cerf}(|\underline{r}_{\ell j}-\underline{\nu}|/2\gamma)}{|\underline{r}_{\ell j}-\underline{\nu}|} + \sum_{\underline{k}}{}' \frac{4\pi}{\Omega}\frac{e^{-\gamma^2 k^2}}{k^2}\left(e^{i\underline{k}\cdot\underline{r}_{\ell j}}-1\right)$$

$$+ \frac{1}{\gamma\sqrt{\pi}} - \sum_{\underline{\nu}}{}' \frac{\mathrm{cerf}(|\underline{\nu}|/2\gamma)}{|\underline{\nu}|}$$

As two particles come very close we find that the limiting value of the expression multiplying $q_\ell q_j$ becomes

$$\frac{\mathrm{cerf}(|\underline{r}_{\ell j}|/2\gamma)}{|\underline{r}_{\ell j}|} + \frac{1}{\gamma\sqrt{\pi}} \to 1/|\underline{r}_{\ell j}| \text{ as expected.}$$

Before proceeding further we shall rewrite V_N as follows:

$$V_N = \frac{1}{2}\sum_j \sum_{\ell\neq j} q_j q_\ell \sum_{\nu} \frac{\mathrm{cerf}(|\underline{r}_{\ell j}-\underline{\nu}|/2\gamma)}{|\underline{r}_{\ell j}-\underline{\nu}|}$$

$$- \frac{1}{2}\sum_j q_j^2 \left[\frac{1}{\gamma\sqrt{\pi}} - \sum_{\underline{\nu}}{}' \frac{\mathrm{cerf}(|\underline{\nu}|/2\gamma)}{|\underline{\nu}|}\right]$$

$$+\frac{1}{2}\sum_{j}\sum_{\ell\neq j} q_j q_\ell {\sum_{\underline{k}}}' \frac{4\pi}{\Omega}\frac{e^{-\gamma^2 k^2}}{k^2}\left(e^{i\underline{k}\cdot\underline{r}_{\ell j}}-1\right)$$

The third term contains $\sum_{j}\sum_{\ell\neq j} q_j q_\ell \left(e^{i\underline{k}\cdot\underline{r}_{\ell j}}-1\right)$ which is identical

with $\left|\sum q_j\, e^{i\underline{k}\cdot\underline{r}_j}\right|^2$ $\left(\text{since} \sum q_j^2 \equiv -\sum_{j}\sum_{\ell\neq j} q_j q_\ell\right)$.

Of the many ways in which this expression has been put to practical use we shall consider the one due to Singer.[30]

We first need to mention the role of γ in the evaluation of V_N. The convergence constant is to be chosen so as to produce the most efficient way of utilizing the summations over $\underline{\nu}$ and $\underline{k}$. For $\gamma \approx 0$ the $\underline{\nu}$ summation and for γ large the $\underline{k}$ summation is the rapidly convergent one. Singer's method consists in the fact that for $\gamma L \sim 5.0$, the $\underline{\nu}$ summation, using the minimum image $\underline{r}_{ij}$ for every pair, can be fairly accurately calculated with the term $\underline{\nu} = 0$ alone; thus we are left with the distance dependent function $\frac{\mathrm{cerf}|\underline{r}_{ij}|/2q}{|\underline{r}_{ij}|}$ and a "reasonable" number of terms need to be kept in evaluating $e^{-\gamma^2 k^2}/k^2$ together with the modulus square of the single particle sum $\sum q_j\, e^{i\underline{k}\cdot\underline{r}_j}$. The constant term of course is $-\frac{1}{2}\sum q_j^2/\gamma\sqrt{\pi}$

We should observe here that the only condition repeatedly used above is $\sum_{j=1}^{N} q_j = 0$. No restriction on the variety of positive and negative charges has been used.

The force on a particle is easily obtained by differentiation. In the Singer method, the $|\underline{r}|$ dependent part gives for $-\partial/\partial x_\ell$,

$$\left(\mathrm{cerf}(|\underline{r}_{\ell j}|/2\gamma) + \frac{2}{\sqrt{\pi}}\, e^{-r_{\ell j}^2/4\gamma^2}\, (|\underline{r}_{\ell j}|/2\gamma)\right) \frac{(x_\ell - x_j)}{|\underline{r}_{\ell j}|^3}$$

, whereas the $\underline{k}$ dependent part gives

$$\frac{1}{2}\sum_{\underline{k}}{}' \frac{4\pi}{\Omega} \frac{e^{-\gamma^2 k^2}}{k^2} \left\{ - ik_x\, q_\ell\, e^{i\underline{k}\cdot\underline{r}_\ell} \sum_j q_j\, e^{-i\underline{k}\cdot\underline{r}} + \text{c.c.} \right\}.$$

In the evaluation of the various $e^{i\underline{k}\cdot\underline{r}_j}$ on the mesh of vectors $\underline{k}$ it is obviously important to avoid repeated calculation of sines and cosines. This is done by the use of recurrence relations.

To summarize, Singer's method consists in (i) choosing a γ so as to reduce the real space part of the Ewald sum to a "short range" central potential, (ii) making full use of recurrence relations i.e. $\exp i(a+b) = \exp(ia) \times \exp(ib)$ in the evaluation of $\exp(i\underline{k}\cdot\underline{r}_j)$ for various $\underline{k}$'s on the mesh and (iii) reducing the double sum

$$\sum_\ell \sum_j e^{i\underline{k}\cdot\underline{r}_{\ell j}}\, q_\ell q_j \text{ to the evaluation of } \left| \sum_\ell q_\ell\, e^{i\underline{k}\cdot\underline{r}_\ell} \right|^2 .$$

EWALD SUM FOR A CELL OF ARBITRARY SHAPE

Let the molecular dynamics cell be constructed out of vectors $a_1\underline{i}$, $a_2\underline{j}$, $a_3\underline{k}$, of length a_1, a_2, a_3 respectively, which are neither all equal in length nor are they mutually orthogonal. The $\underline{k}$ mesh of reciprocal vectors relevant to such a periodically repeating cell is made up of vectors $\underline{k}_1$, $\underline{k}_2$, $\underline{k}_3$, given by $\underline{k}_1 = 2\pi\, a_2 a_3\, (\underline{j} \times \underline{k})/\Omega$ etc. where $\Omega = (a_1 a_2 a_3)\underline{i}\cdot(\underline{j}\times\underline{k})$ is the cell volume. Obviously if $\underline{k} = \lambda\underline{k}_1 + \mu\underline{k}_2 + \nu\underline{k}_3$ and $\underline{r} = x\underline{i} + y\underline{j} + z\underline{k}$ then

$\underline{k}\cdot\underline{r} = 2\pi\left(\frac{\lambda x}{a_1} + \frac{\mu y}{a_2} + \frac{\nu z}{a_3}\right)$. The formula for V_N remains the same as before except that the distance square in real space is not the sum of squares $a_1^2 + a_2^2 + a_3^2$ of the three components but $(a_1\underline{i} + a_2 j + a_3\underline{k})^2$ and $\underline{k}^2$ is $(\lambda\underline{k}_1 + \mu\underline{k}_2 + \nu\underline{k}_3)^2$ which is not $\lambda^2 + \mu^2 + \nu^2$ as it was in the case of a cubic box.[31]

MORE RECENT WORK ON SIMPLIFYING THE COULOMB POTENTIAL PROBLEM

In order to reduce the computational expense involved in using the full fledged V_N and its derivatives S. Brawer[32] has been using the following method in his extensive study of BeF_2 glasses.

Consider the vector $\underline{r}_{\ell j}$ connecting two particles; we first construct the minimum image vector $\underline{r}_{\ell j}$. Now consider the special case when this minimum image $\underline{r}_{\ell j}$ lies along the x axis say (we are considering a cubic box again). Then we have a contribution to V_N from this pair which we will denote by $\psi_{SB}(\underline{r}_{\ell j})$ and similarly a force in the x direction denoted by F_{SB} $(\underline{r}_{\ell j})$. Periodic boundary conditions having been fully built into the Ewald analysis we find that F_{SB} $(|r_{\ell j}| = L/2)$ is exactly zero (for obvious reasons of symmetry).

Thus we have a potential ψ_{SB} and a force F_{SB} which, for pairs of particles with separation along the 3 axes of the molecular dynamics cube, are exactly calculated.

Brawer then uses the same functions to evaluate the potential and the force as a function of the minimum image particle distance $|\underline{r}_{ij}|$, whatever the direction of $\underline{r}_{ij}$ may be. The force is made to act along the direction of the minimum image vector $\underline{r}_{ij}$.

This has an economy factor of two kinds. Firstly the functions ψ_{SB} and F_{SB} can be tabulated and the table used with interpolation schemes of desired sophistication and secondly two particles interact only when their minimum image separation is $\leq L/2$.

T. Soules[33] has suggested a slightly different approach to the reduction, upto $r \leq L/2$, of the forces and potentials between particle pairs, to simple distance dependent functions. If r is the minimum image distance he writes (using $\xi = r/L$)

$$F_r = \frac{q_1 q_2}{L^2}\left(\frac{1}{\xi^2} - \frac{4\xi}{1-\xi}\right) \Big/ \left(1 + \sum_{i=1}^{8} a_i \xi^i\right)$$

$$V_r = \frac{q_1 q_2}{\xi L} \Big/ \left(1 + \sum b_i \xi^i\right) + F(L)$$

It will be very profitable to know what aspects of the structure and dynamics of ionic systems is most sensitive to these simplified versions of the long range Coulomb interaction.

CRYSTAL STRUCTURE OF AgI AND CuI

At low temperatures both AgI and CuI have the Zincblend structure i.e. I^- forms an fcc lattice and the cations are tetrahedrally coordinated to I^-. Cation lattice positions can be obtained by translating the fcc I^- lattice positions by (1/4,1/4,1/4) whereas the (-1/4,-1/4-1/4) positions are unoccupied. The low temperature phase of these two materials is called the γ-phase. As the temperature is raised in both materials there is a crystal structure transformation from γ to β phase which has the Wurtzite structure in which I^- is an hcp lattice and the cations continue to be tetrahedrally coordinated to I^-. The γ-β transition occurs in AgI at 410 K and in CuI at 642 K. In AgI the γ and β phases are invariably found to be mixed together.

In AgI and CuI at 420 K and 680 K, respectively, there is a crystal structure transformation to the α-phase; it is accompanied by a sudden increase in the cation self-diffusion leading to a large dc conductivity.[2] In α-AgI the I^- form a bcc lattice and Ag^+ are mobile whereas in α-CuI the I^- form an fcc lattice and Cu^+ are mobile. Thus the structure of I^- in γ and α-CuI is the same.

The purpose of MD and MC calculations is to understand in microscopic detail, in time and in space, the properties of these superionic conductors.[8,10,13,14,18,19,21,34,35]

SCHEME FOR CONSTRUCTING POTENTIAL FUNCTIONS

The problem of setting up an interaction scheme has been approached[13] from a "phenomenological" point of view rather than from detailed considerations of quantum chemistry which enter into a discussion of electronic distributions around Ag and I ions in a condensed phase. The rules laid down by Pauling imply that much can be understood on the basis of ionic radii deduced from crystal structure data.[15,36] In other words the stability of a crystal is attributed to certain ionic contacts, the sum of the relevant radii are equated to the ionic distances.

The pair potential is taken to be of the form,[13,14]

$$V_{ij} = A_{ij} (\sigma_i + \sigma_j)^{n_{ij}} / r_{ij}^{n_{ij}} + Z_i Z_j e^2 / r_{ij}$$

$$- \frac{1}{2} (\alpha_i Z_j^2 + \alpha_j Z_i^2) \, e^2 / r_{ij}^4 - W_{ij} / r_{ij}^6$$

In AgI, $Z_{Ag} = - Z_I = Z$. α_i are the electronic polarizabilities and $W_{ij} = 3/2\ \alpha_i \alpha_j\ (e_i^{-1} + e_j^{-1})^{-1}$; the e_i are the ionization levels.

In a material such as AgI, it is argued, that the crystal structure is determined principally by I-I and Ag-I contacts. A knowledge of the crystal structure is therefore sufficient to uniquely determine size of ions. In AgI the ionic radii of I^- and Ag^+, σ_I and σ_{Ag} respectively, are defined as,

$$\sigma_I + \sigma_I = \text{I-I nearest neighbor distance}$$

$$\sigma_I + \sigma_{Ag} = \text{I-Ag nearest neighbor distance}$$

If in α-AgI the bcc lattice constant is taken to be 5.06 Å, the assumption that the I-I contact participates in the stability of the crystal gives the I radius $\sigma_I = 5.06 \times \sqrt{3}/4$ or 2.20 Å. Obviously, the unlike ion contact must be involved in crystal stability; assuming the Ag positions to be the tetrahedral sites of the bcc lattice one gets $\sigma_I + \sigma_{Ag} = 5.06 \sqrt{5}/4 = 2.83$ Å. Hence $\sigma_{Ag} = 0.63$ Å. (This should be contrasted with the canonical value of Ag^+ ion radius which is ~ 1 Å).[15,36] In constructing model potentials suitable for a superionic conductor such as AgI one should keep in mind the following general observation. In the periodic table one finds that I^- is among the largest ions (~ 2.2 Å) whereas Ag^+ is among the smallest (~ 1 Å). Since the electronic polarizability α has the dimensions of volume, the polarizability of I^- is much larger than that of Ag^+, we therefore make the simplifying assumption that $\alpha_{Ag} = 0$ in our model.[13,14] Consequently the factor $\alpha_{Ag} Z_I^2$ in the charge-dipole interaction term as well as W_{AgI} and W_{AgAg} in the Van der Waal term vanish. Furthermore, a posteriori it is found that the Van der Waal's interaction term W_{II} does not seem to play an important role; however, it was retained in all our calculations. We have used $\alpha_I = 6.52$ Å^3, $e_I = 3.13$ eV. The number of unknowns is reduced by taking all $A_{ij} = A$ i.e. each "contact" is assumed to contribute the same repulsive energy A. The potential function is further simplified by taking $n_{ij} = n$, i.e., the same repulsive exponent for (I-I), (I-Ag) and (Ag-Ag) terms. We thus need 3 parameters A, n and Z; in principle the low temperature lattice constant, cohesive energy and compressibility should be used to determine the three parameters. In the absence of necessary experimental data required to determine the three constants for AgI and considering the phenomenological manner of constructing the pair potential we have adopted other means of fixing these constants. The analysis of phonon dispersion data of AgI indicated $Z \cong 0.6$ to be an appropriate value. The condition for the energy of the crystal to be a minimum at the lattice constant 5.06 Å gives a relation between n and A. The low temperature compressibility of AgI is known to be $\sim 10^{-11}$ cm^3 erg^{-1}. The value $n = 7$ makes the compressibility of our AgI system of about the same magnitude. We determine $H_{AgAg} = 0.062$, $H_{AgI} = 17.893$ and $H_{II} = 394.934$ where $H_{ij} = A(\sigma_i + \sigma_j)^n$. Here Å is the unit of length and $e^2/\text{Å} = 14.39$ eV is the unit of energy.

The potential functions $V_{ij}(r)$ for α-AgI are shown in Fig. 1. The differences between V_{Ag-Ag} and V_{I-I} at small values of r are simply the manifestation of different sizes of Ag and I ions. The minimum in V_{Ag-I} occurs at about 2.3 Å whereas the Ag-I distance in the solid is 2.8 Å the distance at which the total crystal energy is minimum.

The MD results already reported by us show that this pair potential scheme provides a "working model" for the study of structural and dynamical properties of α-AgI.[13,14]

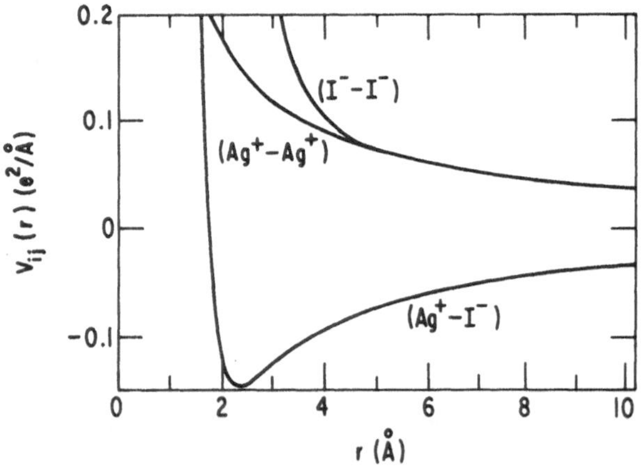

Fig. 1 Effective pair potentials used in MD calculations on α-AgI, in units of $e^2/\text{Å} = 14.39$ eV.

A SUMMARY OF EARLIER RESULTS ON α-AgI

Calculations were performed[13,14] on a 256 particle system in a cubic cell of 20.34 Å with periodic boundary conditions. The mass density of the system was 5.92 g/cc. The long range nature of the Coulomb interaction was properly taken into account by the Ewald method.[30] The calculation was initiated by placing the iodines in a bcc structure and the Ag ions in suitable tetrahedral sites. After sufficiently long initial aging the system was studied between temperatures of 430 and 760 K.

The mean square displacement $\langle[\vec{r}(t+s) - \vec{r}(s)]^2\rangle$ was calculated for iodine and silver ions. The large time behavior of this quantity shows that Ag ions have a large liquid like self diffusion constant, whereas the mean square displacement of iodine ions does not increase with time, implying that iodines have zero diffusion. The asymptotic value of the mean square displacement of the iodine ions is a measure of the Debye-Waller thermal cloud. Figures 2 and 3 show the calculated value of the silver self-diffusion constant, D, and the mean square displacement, B, of the iodines, in good agreement with the experimental results.[37,38]

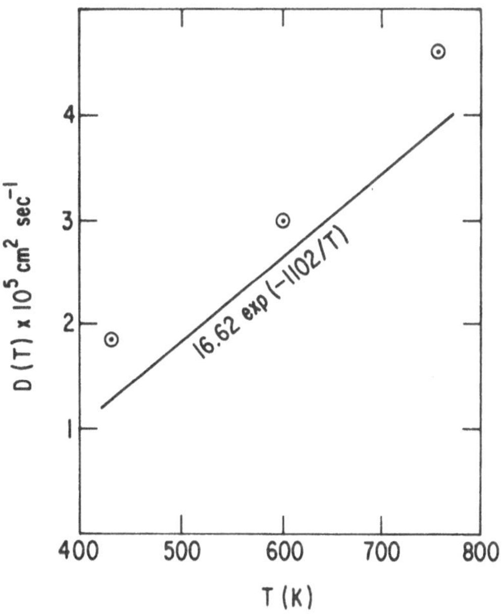

Fig. 2 Temperature dependence of the constant of self diffusion, D(T), for silver in α-AgI. Circles with dot, MD calculation; continuous curve from the experiment of Kvist and Tarneberg (Ref. 37).

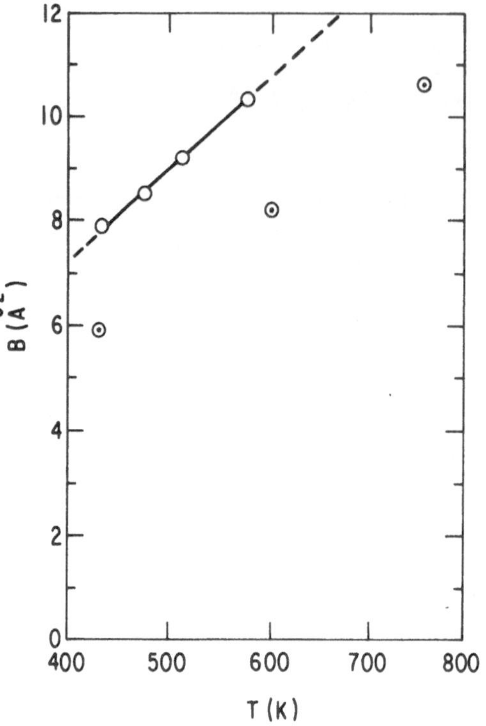

Fig. 3 Temperature dependence of the mean square displacement, B(T), for iodine in α-AgI. Circles with dot, MD calculation; open circles connected by solid line from the experiment of Cava, Reidinger and Wuensch (Ref. 38).

Having established that iodines form a well-defined bcc structure under the influence of the interionic potentials discussed above, it is reasonable to use this structure as a reference to construct a density map of Ag ions in the unit cell. The most interesting region is a (100) face of the bcc lattice. The density map was constructed by monitoring the presence of Ag ions in a plate whose thickness was 1/16th of the lattice constant. The calculated silver density map is in good agreement with the experimentally measured density map of Cava, Reidenger and Wuensch[38] using neutron diffraction. The most important finding of the MD calculation is tha the mid point between two tetrahedral sites is not a maximum of Ag density but a saddle point.[13]

To study the microscopic nature of the mechanism of diffusion, MD trajectories of the system were analyzed using the method of Jacucci and Rahman.[39] We find that the Ag ions diffuse by a jump process with jump frequency 0.34×10^{12} sec^{-1}. Tetrahedral sites on (100) face of a bcc lattice are shown in Fig. 4. Tetrahedral site i has four nearest neighbor "t" sites in [110] directions; j and k are in the plane and the other two are into and out of the plane.

Results of our analysis are shown below:

% of Total Jumps	From	To	Direction
82	"t"	n.n."t"	[110]
9	"t"	n.n.n."t"	[002]
7	"t"	n.n.n.n."t"	[112]

When successive [110] jumps occur, our analysis shows that there is no bias towards a rotation around the "t" sites which form a square on a (100) face of the bcc lattice.

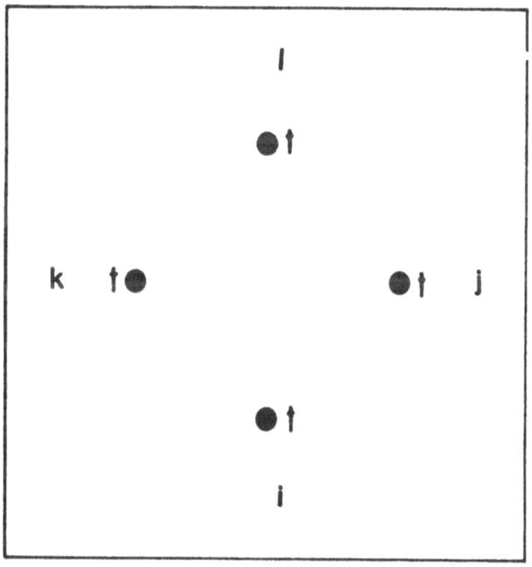

Fig. 4 Four tetrahedral sites on the 100 face of a bcc lattice.

Nature of successive jumps of Ag ions in α-AgI is schematically shown in Fig. 5. For successive [110] jumps there is a bias toward backward jumps in comparison with the other three possibilities which occur with equal probability. We find that among successive [110] jumps 40% in the backward direction and 20% each in the other three directions; in the unbiased case these would be all 25%.

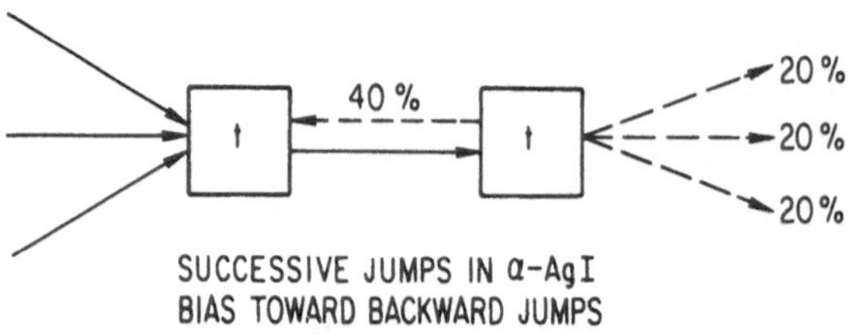

Fig. 5 Schematic diagram of biased "t" to "t" successive jumps in [110] direction. In the unbiased case the probability of jumps in all four directions would be 25%. In α-AgI at 430 K the probability of backward jumps is found to be 40% in the MD calculations.

RESULTS FOR CuI

The results presented in the previous section show that the molecular dynamics calculations performed on our model of α-AgI[13,14] lead to results in good agreement with the experiments.[37,38]

MD calculations were performed on CuI,[14] to test the range of validity of our model, using the same scheme to construct the potential functions as was used in α-AgI. Obviously the crystal structure of CuI was introduced in the scheme to construct the appropriate potential functions. As a consequence all parameters except the H_{ij}'s remained the same as in α-AgI. We determine $H_{CuCu} = 0.01196$, $H_{CuI} = 12.982$ and $H_{II} = 399.578$, in units of $e^2/\text{Å}$.

The calculations were performed on a 216 particle system in a cubic cell of length 18.45 Å with periodic boundary conditions in which, at all temperatures of interest, iodines continue to form an fcc lattice with lattice constant 6.15 Å, the density of the system being 5.44 g/cm^3. The calculations were carried out in the γ and α phases at several temperatures. The vicinity of the superionic transition was studied in detail.

In studying the nature of the superionic transition in CuI at 680 K one assumption is made. The β-phase, which exists in the temperature range 642-680 K, is neglected. In our computer model of CuI the iodine lattice cannot undergo a structural transition and thus remains an fcc lattice at all temperatures of interest, implying that the system goes directly from γ to α-CuI.[14] By neglecting the β-phase we are assuming that the differences between the fcc and hcp structures of iodines, as in γ and β-CuI respectively have no significant effect on the nature of the first order phase transition under consideration.

Various structural and dynamical properties of the computer model of CuI are studied using the MD trajectories. Structural properties include the (Cu-Cu), (Cu-I) and (I-I) pair correlation functions below and above the transition; changes in the intensities of various X-ray reflections below and above the transition. The temperature dependence of mean square displacement of the iodines and the constant of self diffusion of copper have also been studied. In what follows we shall discuss the nature of the superionic transition, the density map of copper ions in the unit cell, and a detailed analysis of copper jumps and diffusion paths.

In the zinc sulfide structure of CuI, the α-phase, there are four copper ions in tetrahedral positions around each iodine. Another set of equivalent tetrahedral sites are unoccupied. Let us denote the occupied sites by (+) and the unoccupied ones by (-). Four occupied (+) sites and four unoccupied (-) sites form a simple cube around each iodine. At very low temperatures the occupation of the (+) sites is unity whereas that of the (-) sites is zero.

In our MD calculations we are dealing with a microcanonical ensemble. The sum of kinetic and potential energies of the system, denoted by E, remains constant as a function of time. The specific heat, C_v, of the systems can be calculated from the fluctuation of the kinetic energy.[40] The temperature dependence of the occupations of (+) and (-) sites, energy per particle, constant of self diffusion of copper, D_{Cu}, and the specific heat are shown in Fig. 6. Below 600 K the occupation of (+) sites is unity and the (-) sites are unoccupied; E(T) has a smooth slope giving rise to a value of $C_v \sim 4\ k_B$ and the D_{Cu} is negligibly small on a scale of 10^{-5} cm^2/sec.

As the temperature of the system is raised E(T), $D_{Cu}(T)$ and $C_v(T)$ increase. This continues, in our computer model, until the temperature reaches about 680 K. In this region which we shall call the pre-transition region E(T) vs. T curve becomes rather steep and C_v increases to a value of 6-7 k_B, at the same time D_{Cu} becomes $\sim 1\text{x}10^{-5}$ cm^2/sec. It is also found that the occupation of (+) sites is about 90% and that of the (-) sites 10%. We found that our system displays substantial diffusion of copper ions before the transition. We show below that this is an order-disorder transition. These results are in qualitative agreement with the experimental observation that before the β-α transition a substantial diffusion of Cu^+ is observed in β-CuI.[2,41]

The transition region in the neighborhood of 700 K in Fig. 6 is marked by vertical dotted lines. When the system is heated

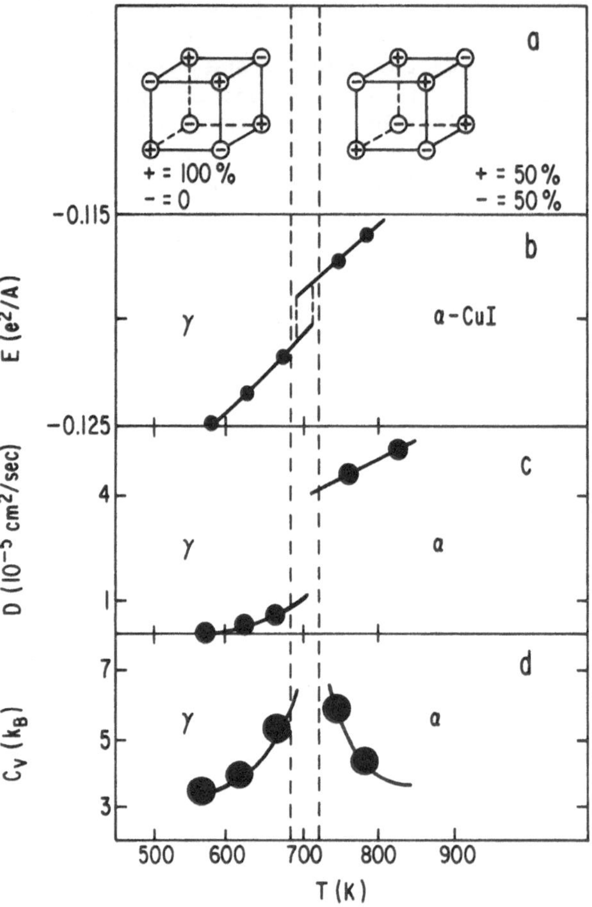

Fig. 6 MD results for the temperature dependence of the various properties of γ and α-CuI. Transition region is marked by two vertical dotted lines around 700 K. (a) Population of (+) and (-) tetrahedral sites in γ and in α-CuI. (b) Total energy (kinetic + potential) per particle in units of $e^2/\text{Å}$ = 14.39 eV. The jump in E(T), from MD calculations, gives a latent heat of 1.33 Kcal/mole. (c) Constant of self diffusion in units of 10^{-5} cm^2/sec. (d) Heat capacity in units of k_B.

above 680 K the energy E(T) increases abruptly. The jump in the energy is the latent heat of transition. The value of C_V in this region is very large; D_{Cu} increases suddenly from 1 to nearly $4x10^{-5}$ cm^2/sec. Above 700 K the occupations of (+) and (-) sites are found to be equal. Above 720 K, E(T), $D_{Cu}(T)$ and $C_V(T)$ show behavior similar to that in the pre-transition region.

The behavior of our system around 700 K can be summarized as follows: The copper ions undergo an order-disorder transformation leading to a specific heat anamoly as well as an abrupt jump in the constant of self diffusion of copper; the system goes through a first-order phase transition, the latent heat involved in the γ to α transition being 1.33 Kcal/mole in reasonable agreement with the experimental value of 2.47 Kcal/mole.[42]

Using a novel experimental technique in which the tracer creation and diffusion are both parts of one integrated procedure, Dejus, Sköld and Graneli[41] have recently measured the constant of self diffusion in CuI. Their experimental results as well as the MD results for the constant of self diffusion of Cu ions are shown in Fig. 7. They find a diffusion constant $4.1x10^{-7}$ cm^2 sec^{-1} in the γ phase at ~ 628 K and $4.6x10^{-6}$ cm^2 sec^{-1} at 661 K in the β phase. The value in the "superionic" α-phase was found to be $1.6x10^{-5}$ cm^2 sec^{-1} at 705 K and $3.5x10^{-5}$ cm^2 sec^{-1} at 834 K. The MD values for CuI are consistently larger but in semi-quantitative agreement with the experimental results. Since Cu, Ag, Au belong to the same column of the periodic table, it is probably justified to assert that $Z_{cu} > Z_{Ag}$ (= 0.6) and Z_{cu} = 0.6 for CuI is an underestimate. A higher value of Z_{cu} will tend to improve agreement with experiment.

The density map of Cu^+ in α-CuI at 700 K is plotted along various paths of interest in the unit cell. This is shown in Fig. 8. Eight tetrahedral sites surrounding an iodine "I" and an adjacent octahedral "O" site of the fcc lattice of iodines are shown in the figure. The Cu^+ density is plotted along the paths ABCDEC marked with the arrows. Starting from a (+) site, marked A, the path goes along 110 to a neighboring (+), marked B, then crossing an "I" site along 111 to (-), marked C, and continuing along 111 crossing an "O" site to (+), marked D. The path then goes along 100 to a (-), marked E, and finally along 110 to a (-) marked C.

As we have remarked earlier, at this temperature the occupation of (+) and (-) sites are equal. On this basis, along with the information presented in Fig. 6, we assert that in our computer model of CuI the $\gamma \rightarrow \alpha$ transition is an order-disorder transition which changes a 100% occupation of one type of site (i.e. a zinc blend structure) into a 50, 50 of both types (i.e. to an "average" fluorite structure). A calculation of the X-ray intensities was carried out at several temperatures below and above the transition using our MD data. The results show how a zinc blend structure transforming into an average fluorite structure manifests itself in the intensities of various reflections which are absent in the zinc blend structure due to lack of inversion symmetry.[43] The results compare favorably with the experimental observations.[42]

Before concluding the discussion on the density map of Cu^+ in α-CuI, we should state the value of Cu^+ density at two important points in the unit cell. When the average density at the tetrahedral sites is normalized to unity, the value at the octahedral "O" site is 0.16 and at the trigonal site (halfway along (+) to (-) 100 path i.e. midway between points D and E) it is 0.04.

Once again we would like to emphasize that structural information such as the density map by itself is not sufficient to deduce the microscopic mechanisms of diffusion and diffusion paths.[2] From the fact that the Cu^+ density at the octahedral site is 0.16 and at the trigonal site only 0.04, it will be erroneous to deduce that the diffusion of Cu^+ in our model occurs purely by jumps across the octahedral sites. To infer the true nature of the diffusion paths one should examine quantities such as the directionality of jumps, total number of jumps in various directions and residence time at a site before the jump in a collective fashion rather than looking at each one of these features individually.

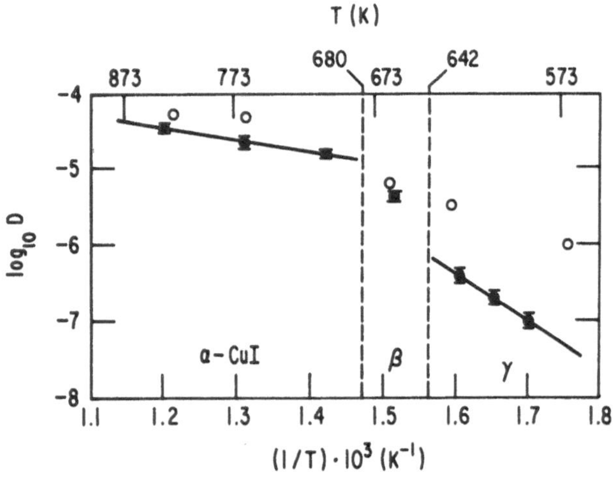

Fig. 7 Temperature dependence of the constant of self diffusion, D(T), of copper in CuI in β and γ-phases. Open circles, MD results in α and γ-phases. Points joined by the continuous line are the experimental results of Dejus, Sköld and Graneli (Ref. 41).

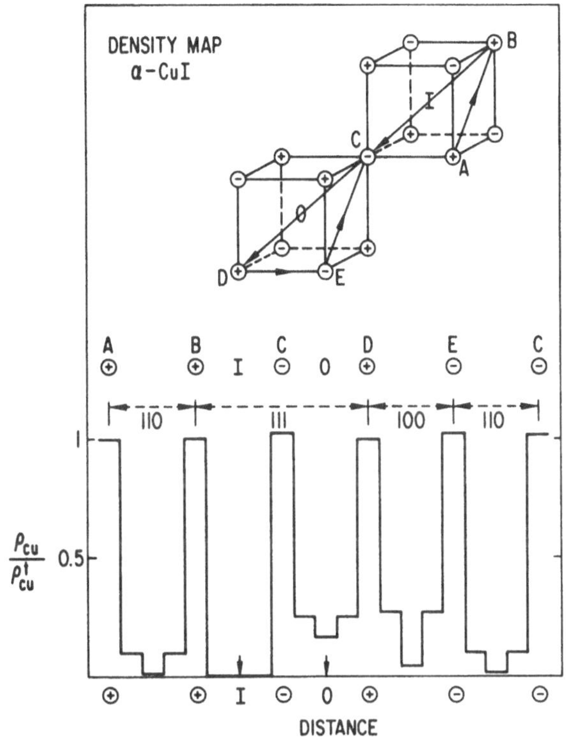

Fig. 8 Density map of copper in α-CuI at 700 K. The paths along which the density is plotted are marked in the inset. "I" refers to the iodine site and "O" to octahedral site. The density is normalized to unity at the tetrahedral sites referred to in the text as (+) and (-) sites.

To investigate the diffusion path of Cu^+ in α-CuI we have used the method of Jacucci and Rahman. The cubic unit cell was divided in 8^3 microcells through which the motion of Cu^+ ions was monitored. This kind of analysis gives in quantitative detail the frequencies of different type of jumps occurring in the system as well as the residence time at various sites. By time averaging the population found in various microcells composing the unit cell we get the density map of Cu^+ as shown in Fig. 8. The jump analysis is given in Table I.

Table I. Analysis of jumps and their directionality for copper ions in α-CuI at 700 K. N is the number of jumps recorded from a microcube to another in a certain direction. τ_R in 10^{-12} sec is the residence time in the old site before the jump.

Type of jump from	to	Number of Jumps N	Residence Time τ_R ps	Jump Direction
(+) (-)	(+) (-)	436	4.9	110
(+) (-)	(-) (+)	1027	3.9	100
(+) (-)	"0" "0"	949	3.9	111
"0" "0"	(+) (-)	947	1.7	111

From the data presented in Table I the following conclusions can be drawn:

(i) As expected, the (+) to (+) or (-) to (-) jumps are less frequent than the other three types shown in the table;

(ii) The jumps along 100 between (+) and (-) sites are as frequent as those between (+) and "0" along 111. This as we have emphasized earlier, could not have been inferred from the density map shown.

It is clear from the above analysis that contrary to what is normally believed,[2] diffusion via octahedral site is not the one and only path for Cu^+ motion in α-CuI.[14]

In the next section we shall present some new results which have been obtained for α-AgI at 450 K, with the potential function given above.

MASS TRANSPORT IN α-AgI AT 450 K

We have recently performed new calculations on a larger system of 500 particles (250 Ag^+ + 250 I^-). In view of the length of the calculation (15000 Δt, $\Delta t = 2.5\times10^{-14}$ s) only one calculation at 450 K, has been completed. Upto now only a few of the quantities of interest have been analyzed.

Already from the detailed analysis[39,44,45] of F^- motion in CaF_2 it was clear that the mean square displacement $\langle[\underline{r}_F(t) - \underline{r}_F(0)]^2\rangle$ by itself showed no qualitative behavior related with the fact that diffusion was taking place with a clearly defined jump process.

The velocity autocorrelation function (vaf) is the second derivative of the mean square displacement:

$$2\langle\underline{V}_i(o)\cdot\underline{V}_i(t)\rangle = \frac{d^2}{dt^2}\langle[\underline{r}_i(t) - \underline{r}_i(o)]^2\rangle$$

or

$$\langle[\underline{r}_i(t) - \underline{r}_i(o)]^2\rangle = 2\int_o^t (t-u)\ \langle\underline{V}_i(o)\cdot\underline{V}_i(u)\rangle\ du$$

Let $\phi(t)$ denote $\langle\underline{V}_i(o)\cdot\underline{V}_i(t)\rangle/\langle\underline{V}_i^2(o)\rangle$. Then

$$D = \frac{k_BT}{M}\int_o^\infty \phi(u)du.$$

where D is the constant of self diffusion and M is the mass of the diffusing ion. In general one can write,

$$\phi(\omega) = \int_o^\infty \phi(t)\ e^{i\omega t}\ dt$$

$$\phi(\omega=o) = MD/k_BT$$

The vaf of the Ag^+ ions is shown in Fig. 9. We note that the constant of self-diffusion which is the integral of the vaf is 1.68×10^{-5} cm^2 s^{-1}, in good agreement with experiment. It is quite remarkable that the Ag^+ velocity auto correlation function obtained in α-AgI has all the familiar characteristics[46] of the same function for Argon atoms in liquid Argon at the triple point! In other words, the characteristics of the jumps of Ag^+ ions in α-AgI do not reveal themselves in the vaf. Attempts to understand the phenomena related with superionic conduction in AgI through the velocity autocorrelation function are not therefore useful.[47,48]

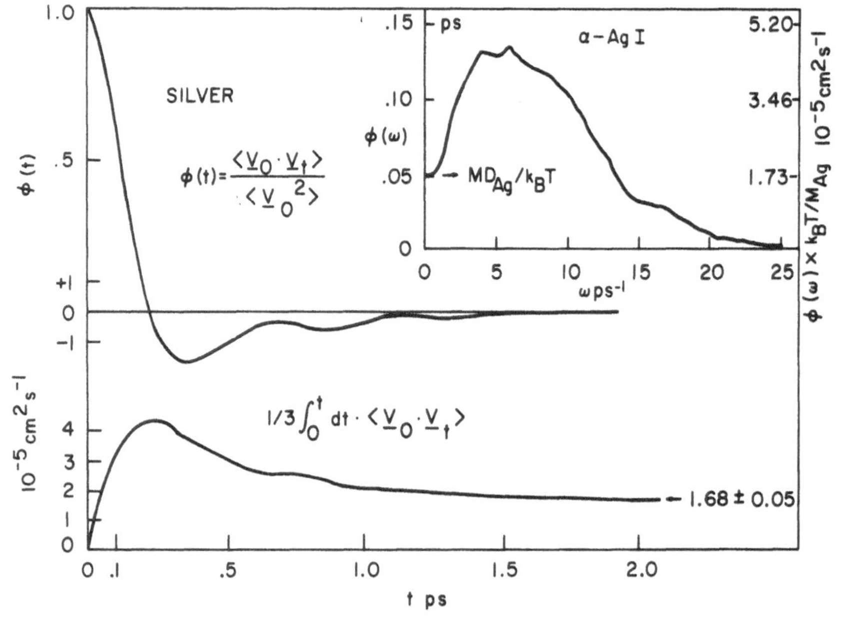

Fig. 9 The normalized velocity autocorrelation function (vaf), $\phi(t)$, for silver in α-AgI at 450 K. The Fourier transform multiplied by k_BT/M_{Ag} is shown in the inset. Integral of the vaf is shown in the lower part of the figure. The asymptotic value of the integral gives the constant of self diffusion for silver, 1.68×10^{-5} cm^2/sec. Note that the vaf for silver has remarkable similarity to the vaf for the monoatomic simple liquid like Argon. Hence, no information about the actual diffusion mechanism of silver in α-AgI can be drawn from the vaf or from its frequency spectrum shown in the figure.

In Fig. 9 we have shown not only the silver vaf but also its Fourier transform and the manner in which its time integral converges to the final value which gives the constant of self diffusion for silver, $D = 1.68 \times 10^{-5}$ cm^2/sec.[13,37] We again note the lack of any definite structure in the Fourier transform as is the case in monotonic fluids.[46]

The velocity auto correlation for Iodine is shown in Fig. 10 together with its integral and its Fourier transform. It is clear that the I ions do not diffuse and also that the spectrum of their vibratory motion around the lattice sites has no resemblance with the spectrum of the density of states usually associated with a harmonically vibrating bcc lattice.

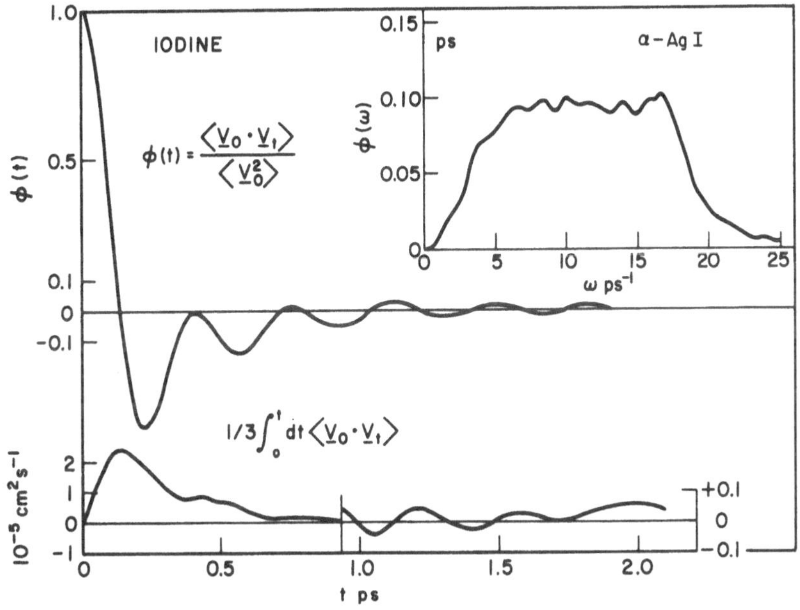

Fig. 10 The normalized velocity autocorrelation function, $\phi(t)$, for iodine in α-AgI. The Fourier transform $\phi(\omega)$ and time integral are also shown. The asymptotic value of the time integral shows zero diffusion for iodine.

It needs to be emphasized once again that the velocity autocorrelation function and hence its spectrum is related to the long wave length ($\underline{k} \to 0$) limit of the behavior of the function $F_s(\underline{k},t) = \langle \exp i\underline{k}\cdot[\underline{r}_i(t) - \underline{r}_i(0)]\rangle$[49,50] and if the motion occurs with jumps of length 10^{-8} cm in specific directions one cannot observe such details except through the behavior of $F_s(\underline{k},t)$ with vectors $\underline{k}$ of the order 1 to 3 Å^{-1} in length. Averaging over directions of $\underline{k}$ will again lead to a "mixing up" of information about the jumps; this mixing is impossible to disentangle. A clear and detailed analysis of these notions has already been made for CaF_2 by Jacucci and Rahman.[39,44]

It should be remarked here that neutron inelastic scattering experiments in superionic conductors should be made on single crystal samples and the spectrum of energy loss and gain should be obtained at constant values of the momentum transfer vector. We recall that experiments[51] on metal hydrides have provided data which is worth a detailed analysis because the experiments were made satisfying the above constraints.

CHARGE TRANSPORT IN α-AgI

For the calculation of conductivity in an MD system it is necessary to correlate the charge current fluctuations. We use the basic Kubo relation,[49,50]

$$\sigma(\omega) = \frac{1}{3\Omega kT} \int_0^\infty \langle \underline{J}_0 \cdot \underline{J}_t \rangle \, e^{i\omega t} \, dt \quad .$$

σ is the conductivity, Ω the volume and $\underline{J}_t$ is the charge current, i.e. $\sum_j q_j \underline{v}_j(t)$ where q_j is the charge of the particle with velocity $\underline{v}_j(t)$ at time t. In fact if we define $\underline{M}_t = \sum q_j \underline{r}_j(t)$ then $\underline{J}_t = \dot{\underline{M}}_t$ and

$$2\langle \underline{J}_0 \cdot \underline{J}_t \rangle = \frac{d^2}{dt^2} \langle (\underline{M}_t - \underline{M}_0)^2 \rangle$$

Let $\Phi(t)$ denote $\langle \underline{J}_0 \cdot \underline{J}_t \rangle / \langle \underline{J}_0^2 \rangle$. Then

$$\sigma(\omega) = \frac{\langle \underline{J}_0^2 \rangle}{3\Omega kT} \int_0^\infty \Phi(t) e^{i\omega t} \, dt \quad .$$

But $\langle \underline{J}_0^2 \rangle = 3k_B T \sum_j (q_j^2/M_j)$; in the case of AgI it is $q^2 \; 3k_B T/\mu \, N_{Ag}$ where $\mu^{-1} = M_{Ag}^{-1} + M_I^{-1}$ and since all the silver ions are mobile in α-AgI, $N_{Ag} = N_I = N$.

Hence, for AgI,

$$\sigma(\omega) = q^2 \frac{N_{Ag}}{\Omega} \mu^{-1} \int_0^\infty \phi(t)\, e^{i\omega t}\, dt$$

It is usual to express $\sigma(\omega=0)$ as a diffusion constant. This conversion (the Nernst-Einstein expression) at any ω can be written as

$$D_\sigma(\omega) = \frac{\Omega kT}{q^2 N_{Ag}} \sigma(\omega)$$

$$= \frac{kT}{\mu} \int_0^\infty \phi(t)\, e^{i\omega t}\, dt$$

in complete analogy with the equation for particle (i.e. tracer) diffusion.

The function $\phi(t)$ is shown in Fig. 11 and $D_\sigma(\omega)$ in Fig. 12.

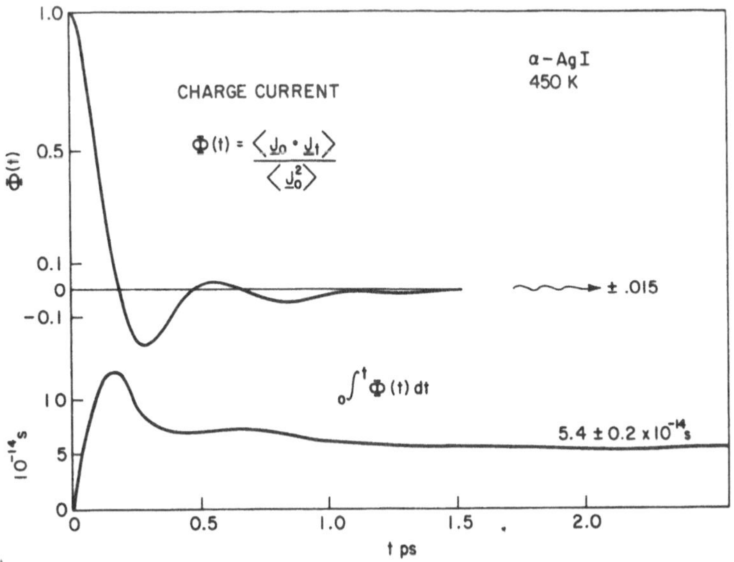

Fig. 11 The normalized current-current correlation, $\Phi(t)$, in α-AgI at 450 K. The time integral is shown in the lower part of the figure. The asymptotic value of the time integral, 5.4×10^{-4} sec. with proper multiplying factors is the zero frequency electrical conductivity.

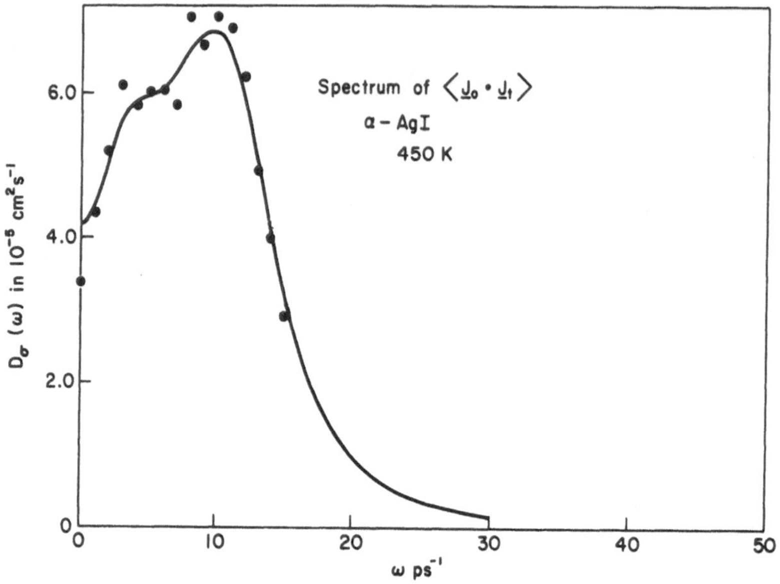

Fig. 12 The Fourier transform of the normalized current-current correlation function, shown in Figure 11. Solid dots were obtained when the correlation function was truncated in time at 3ps; when a smoothing function was used one obtained the continuous curve. The error bar on the dots is about 0.5 in the units shown on the ordinate.

The so called Haven's ratio[52] is then

$$D_{Ag}/D_{\sigma}(0) = 1.68/3.4 \text{ or } \sim 0.5.$$

The fact that the experimental value[37,53] of this ratio is ∿ 0.6 again emphasizes the correctness of the model of α-AgI that we are dealing with. The MD value of ∿ 0.5 for the Haven's ratio is also in good agreement with the theoretical estimate.[54]

STRUCTURAL TRANSFORMATIONS IN AgI

As we discussed earlier, AgI undergoes a number of interesting structural transitions as a function of pressure (or density) and temperature.[2,16,55-60] In the previous section we discussed in detail the structural and dynamical properties of α-AgI and CuI. In this section we shall report on our studies of structural phase transitions in AgI. It should be remarked here that within the context of MD or MC technique structural transitions are among the most difficult to study.

The results described in the previous sections were obtained using "traditional MD" or "old MD" technique by which we mean that in the MD calculation the system was confined in a fixed volume of constant shape with periodic boundary conditions.[22]

The iodine lattice, confined in a cubic cell and given a bcc structure, cannot undergo a structural transition and transform e.g. into an fcc structure when the "old MD" technique is used.

Consider N-particles in an MD cell of cubic shape with sides of length a and volume $\Omega = a^3$. In "old MD" technique the volume and thus the number density $\rho = N/\Omega$ remains constant as a function of time. Andersen[61] has generalized the MD technique to include the change in the volume of the cell, i.e., in his method $\Omega_{cube}(t)$ changes with time. It should be remarked, however, that the bcc lattice in a cubic cell cannot transform into another lattice structure even though the volume of the cubic cell can change with time. Therefore, for a study of the structural phase transitions it is essential to incorporate changes as a function of time in the volume as well as the shape of the cell[20] within the framework of molecular dynamics technique. One of the reasons why AgI is such an interesting material is that it undergoes a sequence of crystal structure transformations. The most remarkable feature in the behavior of AgI is the structural difference between γ and β-AgI on the one hand and α-AgI on the other. The former have a close packed iodine structure while in α-AgI the iodines form a bcc lattice. As the pressure is increased the γ-AgI transforms into a rocksalt structure.[55,57,58] One important difference between these two structures is that while iodine forms an fcc lattice in both phases, around each iodine in the nearest neighbor positions there are 4 Ag ions in γ-AgI and 6 in the rocksalt structure.

It has now become possible to study this structural change by computer simulation. Recent work of Parrinello and Rahman[20] has shown that, by using an appropriate Lagrangian one can set up a molecular dynamics calculation in which both the volume and the shape of the periodically repeating cell change with time. The direction of the change is dictated by the difference between the externally applied stresses and those generated internally because of the velocities of the particles and the forces acting between them. Parrinello and Rahman[20] have demonstrated the utility of these new dynamical equations by applying them to simple monatomic systems. They showed that if a system of Lennard-Jones atoms is given a bcc structure then under suitable density, temperature conditions the dynamical equations themselves make the system change its structure to a close packed one. The reverse occurs for a system which is supposed to simulate solid rubidium.

A brief comparison between the old and new MD techniques is given in the following table.

Table 2

Old MD	New MD
Constant cell volume	Variable cell shape and cell volume
Number of particles constant	Number of particles constant
Density constant, pressure varies with time	Pressure constant, density varies with time
Energy E = constant	Enthalpy $H = E + p\Omega$ = constant
3N variables	3N + 9 variables

The additional 9 dynamical variables describe the time dependence of the shape of the MD cell. The change in shape of the MD cell is determined by the difference between internal (kinetic + potential) stress tensor and the externally applied stress.

Parrinello, Rahman and Vashishta[21] have used this new technique to study structural transformation in AgI. The aim of this calculation was to invent a simple potential function scheme for AgI which will make the iodines adopt a bcc structure at elevated temperatures with the silver ions jumping between tetrahedral sites in the manner discussed in the previous section. On cooling the system the interaction potentials all on their own should modify the bcc structure into a close packed one while at the same time the silver ions should settle down into a four coordinated non-diffusive configuration with respect to the iodines.

It is important to understand the following geometrical facts when studying the bcc to fcc transformation of the iodine lattice in α to γ phase change of AgI using the "new MD" technique in which the shape and size of the MD cell change with time. As shown in Fig. 13, consider a bcc lattice in a cube whose side is of length a. The plane ABCD passing through the face diagonal, is a rectangle with sides $\sqrt{2}a$ and a. It has an atom at the center and four atoms at the corners. If the cube is stretched along the edge BC such that A'B' = B'C', then A'B'C'D' is indeed a square and the (100) face of an fcc lattice with lattice constant $\sqrt{2}b$. The main point of this illustration is to convey the fact that a body centered tetragonal lattice with edges, 1, 1, $\sqrt{2}$ is an fcc structure and inversely a tetragonal face centered structure with edges $\sqrt{2}$, $\sqrt{2}$, 1 is a bcc structure. hcp and simple cubic structures can also be generated out of a body centered tetragonal structure by appropriate changes of lengths, angles and the position of the body center.

Because of the success in the study of α-AgI using the pair potentials already outlined above within the framework of "old MD" techniques it was hoped that the above mentioned structural transformations should be produced in the simulation with only minor modification of those potentials. Due to the requirement of the old MD technique, in the study of α-AgI discussed in the previous sections the iodine lattice was given a bcc structure which fitted perfectly in the cubic MD cell. For this reason the parameters

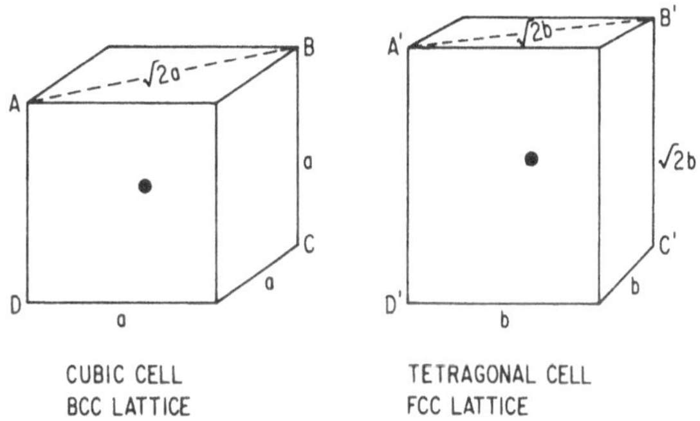

Fig. 13 Schematic representation to show that a body centered tetragonal lattice with edges, 1, 1, $\sqrt{2}$ is an fcc structure.

σ_{Ag} = 0.63 and σ_I = 2.2 Å occurring in the potential functions which were used for the study of α-AgI were determined using the crystal structure of α-AgI. To get a reasonable value of the compressibility the value for the exponents in the repulsive term were taken to be $n_{II} = n_{AgI} = n_{AgAg} = n = 7$. Since with the new MD technique we can study the structural phase transition of the iodine lattice, we are not confined to determine the constants in the potential function from the structure of α-AgI; instead we have used the low temperature structure of AgI to determine the constants. The potential functions were refined by using different values of the exponents n_{ij}. Using the experience with ionic solids we kept n_{II} = 7 but chose a higher value for the (Ag,Ag) repulsive term, n_{AgAg} = 11. A mean of the two values was taken for the (Ag-I) term, n_{AgI} = 9. The value of charge, Z = 0.6, was kept the same as before. We shall call the potential function with these constants as the "modified potential." We show below the modified potential in full detail.

$$V_{AgAg} = H_{AgAg}/r^{11} + 0.36/r$$

$$V_{AgI} = H_{AgI}/r^{9} - 0.36/r - 1.1736/r^{4}$$

$$V_{II} = H_{II}/r^{7} + 0.36/r - 2.3472/r^{4} - 6.9331/r^{6}$$

where

$$H_{AgAg} = A(\sigma_{Ag} + \sigma_{Ag})^{11} = .014804,$$

$$H_{AgI} = A\,(\sigma_{Ag} + \sigma_{I})^{9} = 114.48$$

$$H_{II} = A\,(\sigma_{I} + \sigma_{I})^{7} = 446.64,$$

$$A = 0.010248$$

The length is measured in Å units and energy in units of $e^2/\text{Å}$ (= 14.39 eV).

With the new MD technique the standard system for the purpose of studying structural phase transition of I^- lattice as well as the order-disorder transformation of Ag^+ ions is an N = 500 ion (250 I^- + 250 Ag^+) neutral system obeying the 3N+9 equations of motion.[21] Note that in the traditional MD technique, for an N = 500 particle system in a non-varying cubic cell one can study <u>only</u> the properties of α-AgI and none of the other structures of AgI. This is due to the fact that a cubic cell cannot accommodate 250 I^- ions in perfect fcc or hcp structure.

As a starting point we have taken a well equilibrated 500 particle α-AgI system at T = 700 K in a cubic cell from our previously reported studies. This system will be referred to as "old: α-AgI: T = 700."

First we establish the fact that the modified potential functions with the new MD technique at T = 700 K leads to a satisfactory description of the α-AgI system. For a calculation extending over 5×10^{-11} sec (~ 2500 integration steps of MD) we find a system with normal α-AgI behavior. The I^- ions continue to form a stable bcc lattice with an apparent n.n. coordination of 14, as expected in a heated bcc lattice : the first two shells containing 8 and 6 particles respectively merge into single shell of 14. The Ag^+ diffuse with a constant of self-diffusion $D_{Ag} = 4x10^{-5}$ cm^2/sec. The pair correlation function for (Ag-I) shows a n.n. coordination of 4. This is as has been found in our calculations using old MD technique with constant cubic cell, as discussed in earlier sections.[13,14]

Having confirmed that the "modified potential" describes α-AgI rather well, we proceeded to study the structural phase transitions in AgI as a function of temperature and density.

A well equilibrated 500 particle α-AgI system in a "cube" at 700 K with an apparent n.n. (I-I) coordination of 14, (Ag-Ag) coordination of 4 and having a constant of self diffusion $D_{Ag} = 4x10^{-5}$ cm^2/sec was cooled to 350K. Structural and dynamical behavior of iodines and silvers was monitored as a function of time.

Only after 1,000 time steps ($2x10^{-11}$ sec) it became clear that n.n. (I-I) coordination was changing from 14 toward 12 indicating a structural change of the iodine lattice from bcc toward fcc or hcp. The n.n. (Ag-Ag) coordination remained unchanged from the value 4. The system was well equilibrated for 2,500 time steps ($5x10^{-11}$ sec) and no further changes in the n.n. (I-I) or (Ag-Ag) coordinations were observed as a function of time. The constant of self-diffusion for both silver and iodine was found to be zero showing that both types of ions oscillate around stable mean positions. The mean square displacement was found to be $B_{Ag} = 6.3$ Å^2 and $B_I = 4.0$ Å^2. The partial pair correlation functions for (Ag-I) and (I-I) are shown in Fig. 14. The nearest neighbor iodine and silver coordinations clearly indicate that the α-AgI system has transformed into a Zinc blend or Wurtzite structure. Upon examining the MD cell structure it was found that the cubic shape had transformed into a non-cubic cell. From the (I-I) pair correlation function in Fig. 14 it cannot be determined conclusively whether the iodine lattice is fcc or hcp, though the structure around 10 Å indicates an hcp structure. To sharpen the peaks in the (I-I) pair correlation function, the system was quenched. The results are shown in Fig. 15.

Integrating the (I-I) pair correlation upto the minimum at 9.8 Å (Fig. 15) the value obtained for the coordination number is 55.6. In an hcp structure this coordination is 56 and it is 54 in an fcc arrangement. The discrepancy between 55.6 and the ideal value 56 is probably due to stacking faults in the close packed structure. Such defects have been found in similar calculations by Parrinello and Rahman.[20] Thus we conclude that the α→β transition was successfully achieved upon cooling α-AgI.

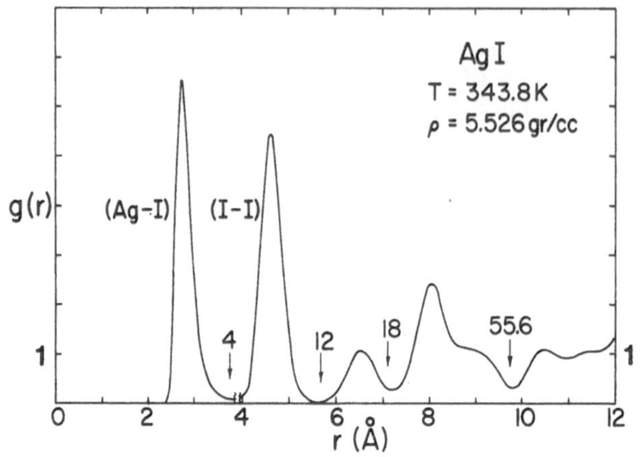

Fig. 14 Partial pair correlation functions for AgI at 343.8 K, the system having been cooled from 700 K. On the left, the peak is the first peak of the (Ag-I) pair correlation; the number of silvers under this peak is 4 as shown by the arrow. The complete I-I pair correlation is also shown; the first peak contains 12 neighbors as indicated by the arrow. Note the arrow at 9.75 Å indicating a total coordination of 55.6 iodines. For a perfect fcc and hcp lattice the number is 54 and 56, respectively.

The $\beta \rightarrow \alpha$ transition was studied by heating to 495 K the β-AgI system obtained as explained above. After ~ 1000 time steps the n.n. (I-I) coordination started to be more than 12. A well thermalized system at 495 K, Fig. 16, clearly shows a n.n. (I-I) coordination of 14 : the characteristic value for a heated bcc lattice. Silver ions were found to be diffusing with $D_{Ag} = 2 \times 10^{-5}$ cm^2/sec and n.n. (Ag-I) coordination was determined to be tetrahedral. To confirm the structure of the iodine lattice the system was quenched. This led to a clear identification of the first two shells of the bcc lattice each containing 8 and 6 particles, respectively.

Using several MD calculations we studied the $\alpha \rightarrow \beta$ and $\beta \rightarrow \alpha$ transition; these calculations indicated that the transition temperature lies between 472 K and 495 K. This is in satisfactory agreement with the experimental value of 420 K.

The principal shortcoming of the potential scheme is that it does not produce the experimentally observed density increase in $\beta \rightarrow \alpha$ transition. All the structural and dynamical properties of AgI, however, are in semi-quantitative accord with experiments.

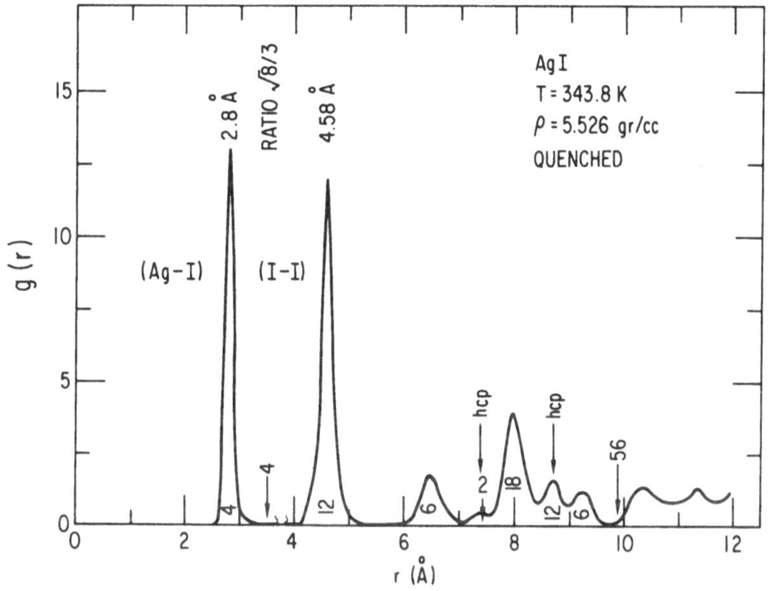

Fig. 15 Partial pair correlation functions for AgI at 343.8 K on quenching to a very low temperature to eliminate thermal vibration. On the left, the (Ag-I) peak at 2.8 Å is sharpened and shows a coordination of 4. The (I-I) pair correlation shows much more structure upon quenching. The first peak at 4.6 Å shows a coordination of 12. The ratio between (I-I) and (Ag-I) n.n. distances is $\sqrt{8/3}$. Also, clearly seen are the peaks containing 6, 2, 18, 12 and 6 particles. There are six peaks up to 9.75 Å containing a total of 56 particles; two peaks marked hcp, contain 2 and 12 particles. This shows that iodines form an hcp lattice and that the pair correlation are for β-AgI.

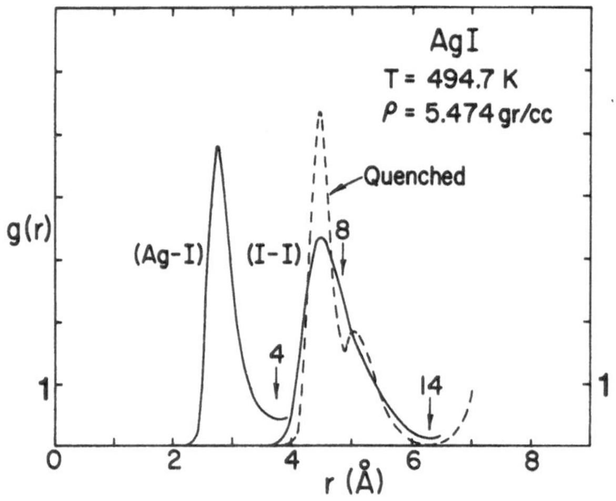

Fig. 16 Pair correlation functions for AgI at 494.7 K obtained by heating from β-AgI at 343.8 K. On the left, continuous curve, the first peak of (Ag-I) pair correlation containing 4 particles. The peak in (I-I) correlation, continuous curve, shows an apparent coordination of 14. This is characteristic of a thermally agitated bcc lattice. Upon quenching the system, (I-I) correlation, dashed curve, shows two peaks containing 8 and 6 particles respectively. This confirms that iodines form a bcc lattice and that the figure describes the pair correlations of α-AgI.

One important quantity which can be accurately measured by EXAFS is the first peak of the (Ag-I) pair correlation function. In Fig. 17 we compare our old and new MD results with the experimental results of Hayes and Boyce.[62,63] The agreement between the EXAFS and the new MD results is indeed very good. To account for the discrepancy of ∿ .05 Å, it is probably necessary to make

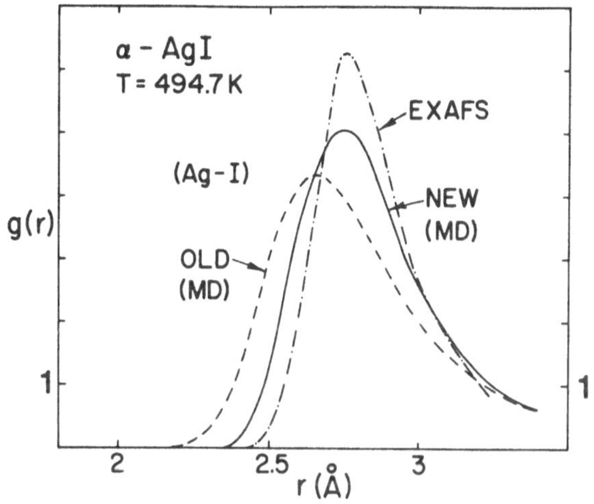

Fig. 17 The first peak of the (Ag-I) pair correlation function in α-AgI. "Old MD" results, dashed curve, continuous curve "New MD" results. EXAFS results of Hayes and Boyce, dash dot curve. It is clear that "New MD" results are in better agreement with the experiment.

It should be emphasized that the potential function was not modified to improve agreement with the EXAFS data. Rather, it was

further MD and EXAFS investigations. It is obvious that the modified potential function which has different exponents (n_{II} = 11, n_{Ag-I} = 9 and n_{AgAg} = 7) to describe the (I-I), (Ag-I) and (Ag-Ag) repulsive terms gives better agreement with the EXAFS results compared to the previous potential ($n = n_{II} = n_{AgI} = n_{AgAg} = 7$).

modified for use with the new MD technique to produce the correct sequence of very intricate structural transitions in AgI as a function of temperature. It is indeed gratifying that as a byproduct of the modification of the potential function, the new (Ag-I) pair correlation function is in better agreement with the experimental EXAFS data.

To reiterate this point, we should remark that the old potential function can be modified in a variety of ways to improve the agreement with the EXAFS data, however, such a modified potential would not necessarily produce correct structural phase transitions in AgI. In summary, while we feel that EXAFS data provides a sensitive check of MD results and in turn of the quality of the potential functions used to simulate the system, the reverse is not necessarily true; the EXAFS data alone cannot be used to determine the pair potential functions. Structural phase transitions are a much more severe test of the quality of the effective pair potential functions.

In the past several years there has been a great deal of theoretical activity with the aim of understanding the structural and dynamical properties of superionic conductors such as AgI and CuI[3-8,12,64-68]; since a great deal of MD results are now available on AgI and CuI based on a rather simple model of effective pair potential functions, it may be worthwhile for theorists to make renewed efforts toward a quantitative understanding of superionic conductors.

ACKNOWLEDGEMENTS

The authors would like to thank M. Parrinello and R. K. Kalia for stimulating discussions and Bonnie Russell, Janice Shelby and Steve Smith for great patience and assistance in preparation of the manuscript.

REFERENCES

1. C. Tubandt and E. Lorenz, Z. Physik, Chem. 87, 513 (1914).

2. K. Funke, Prog. Solid State Chem. 11, 345 (1976).

3. G.D. Mahan and W.L. Roth (Eds.), Superionic Conductors, Plenum, New York, 1976.

4. S. Geller (Ed.), Solid Electrolytes, Springer-Verlag, New York, 1977.

5. P. Hagenmuller and W. Van Gool (Eds.), Solid Electrolytes, Academic Press, New York, 1978.

6. M.B. Salamon (Ed.), Physics of Superionic Conductors, Springer-Verlag, New York, 1979.

7. J.B. Boyce and B.A. Huberman, Physics Reports, 51, 190 (1979).

8. P. Vashishta, J.N. Mundy and G.K. Shenoy (Eds.), Fast Ion Transport in Solids, Elsevier North-Holland, New York, 1979.

9. W.F. Flygare and R.A. Huggins, J. Phys. Chem. Solids 34, 1119 (1973).

10. W. Schommers, Phys. Rev. Letters 38, 1536 (1977).

11. W. Buhrer, R.M. Nicklow and P. Bruesch, Phys. Rev. B17, 3362 (1978).

12. W. Andreoni and J.C. Phillips, Phys. Rev. B , to be published.

13. P. Vashishta and A. Rahman, Phys. Rev. Letters 40, 1337 (1978).

14. P. Vashishta and A. Rahman, in Fast Ion Transport in Solids (Eds., P. Vashishta, J.N. Mundy and G.K. Shenoy), Elsevier North-Holland, 1979, p. 527.

15. L. Pauling, The Nature of the Chemical Bond, Cornell University Press, Ithaca, New York, 1960.

16. L.W. Strock, Z. Physik. Chemie B25, 411 (1934); B31, 132 (1936).

17. G. Burley, J. Phys. Chem. Solids 25, 629 (1964).

18. P. Vashishta and A. Rahman, in Third International Meeting on Solid Electrolytes-Solid State Ionics and Galvanic Cells, Sept. 15-19, 1980, Tokyo Japan, Extended Abstracts, p. 187.

19. P. Vashishta, Bull. Am. Phys. Soc. 26, 372 (1981).

20. M. Parrinello and A. Rahman, Phys. Rev. Letters 45, 1196 (1980).

21. M. Parrinello, A. Rahman and P. Vashishta, to be published.

22. A. Rahman, in Correlation Functions and Quasiparticle Interactions in Condensed Matter (Ed. J. Woods Halley, Plenum, 1977), p. 417.

23. L. Verlet, Phys. Rev. 159, 98 (1967).

24. C.W. Gear, ANL Report No. 7126, Argonne National Laboratory (1966); Numerical Initial Value Problem in Ordinary Differential Equations, Prentice Hall, Englewood Cliffs, N.J., 1971.

25. A. Rahman and F. Stillinger, J. Chem. Phys. 55, 3336 (1971).

26. J.D. Weeks, D. Chandler and H.C. Andersen, J. Chem. Phys. 54, 5237 (1971); 55, 5422 (1971).

27. R.W. Hockney, S.P. Goel and J.W. Eastwood, Chem. Phys. Letters 21, 589 (1973).

28. M. Mandell, J. Statistical Physics 15, 299 (1976).

29. S.W. de Leeuw, J.W. Perram and E.R. Smith, Proc. R. Soc. Lond. A373, 27 (1980); A373, 57 (1980).

30. K. Singer's method has been explained in detail in, M.J.L. Sangster and M. Dixon, Adv. in Physics 25, 247 (1976).

31. M. Parrinello and A. Rahman have evaluated Ewald Sums for arbitrary shaped MD cells in their study of polymorphic transitions in alkali halides using the method of Ref. 20.

32. S. Brawer (Private Communication). Also see J. Chem. Phys. 72, 4264 (1980).

33. T. Soules, J. Chem. Phys. 72, 6314 (1980).

34. Y. Hiwatari and A. Ueda, in Third International Meeting on Solid Electrolytes - Solid State Ionics and Galvanic Cells, Sept. 15-19, 1980, Tokyo, Japan, Extend Abstracts, p. 202; see also A. Fukumoto, Y. Hiwatari and A. Ueda, p. 205.

35. Y. Tsuchiya, S. Tamaki and Y. Waseda, J. Phys. C: Solid State Phys. 12, 5361 (1979); R.J. Cava and D.B. McWhan, Phys. Rev. Letters 15, 2046 (1980).

36. C. Kittel, Introduction to Solid State Physics, Wiley, New York, 1971.

37. A. Kvist and R. Tarneberg, Z. Naturforsch. 25A, 257 (1970).

38. R. J. Cava, F. Reidinger and B. J. Wuensch, Solid State Commun. 24, 411 (1977).

39. G. Jacucci and A. Rahman, J. Chem. Phys. 69, 4117 (1978).

40. L. Lebowitz, J.K. Percus and L. Verlet, Phys. Rev. 153, 250 (1967); see also W.C. Kerr, Phys. Rev. B19, 5773 (1979).

41. R.K. Dejus, K. Sköld and B. Graneli, Solid State Ionics, 1, 327 (1980).

42. K. Miyake, S. Hoshino and T. Takenaka, J. Phys. Soc. Japan 7, 19 (1952); Errata 7, 339 (1952).

43. A. Rahman and P. Vashishta, using MD trajectories for CuI the intensities of X-ray lines is calculated. The results are in good agreement with Ref. 42.

44. A. Rahman, J. Chem. Phys. 65, 4585 (1976); M. Dixon and M. Gillan, J. Phys. C 11, L165 (1978).

45. A. Rahman, in Fast Ion Transport in Solids (Eds. P. Vashishta, J.N. Mundy and G.K. Shenoy), Elsevier North-Holland, 1979, p. 643.

46. A. Rahman, Phys. Rev. 136, A405 (1964).

47. M. O'Keeffee, in Superionic Conductors (Eds. G.D. Mahan and W.L. Roth), Plenum, New York, 1976, p. 101.

48. K. Funke in, Festkorperprobleme (Advances in Solid State Physics), XX, 1 (Ed. J. Treusch), Vieweg, Braumschweig, 1980.

49. P.A. Eaglestaff, An Introduction to the Liquid State, Academic Press, London, 1967.

50. J.P. Hansen and I.R. McDonald, Theory of Simple Liquids, Academic Press, London, 1976.

51. A. Rahman, K. Sköld, C. Pelizzari, S.K. Sinha and H.E. Flotow, Phys. Rev. B14, 3630 (1976); C.J. Glinka, J.M. Rowe, J.J. Rush, A. Rahman, S.K. Sinha and H.E. Flotow, Phys. Rev. B17, 488 (1978).

52. K. Kompaan and Y. Haven, Trans. Faraday Soc. 52, 786 (1956); 54, 1498 (1958).

53. P. Jordan and M. Pochon, Helv. Phys. Acta 30, 33 (1957).

54. I. Yokota, J. Phys. Soc. Japan 21, 420 (1966); H. O. Kazaki, J. Phys. Soc. Japan 23, 355 (1967).

55. B.L. Davis and L.H. Adams, Science 146, 519 (1964).

56. W.A. Bassett, T. Takahashi, Am. Geophys. Union. Trans. 45, 121 (1964).

57. P.W. Bridgman, Proc. Am. Acad. Arts Sci. 51, 55 (1915).

58. G.J. Piermarini and C.E. Weir, J. Res. Natl. Bur. Stds. 66A, 325 (1962).

59. B.M. Riggleman and H.G. Drickamer, J. Chem. Phys. 38, 2721 (1963).

60. B.-E. Mellander, A. Lunden and M. Friesel, in International Conference on Fast Ionic Transport in Solids, May 18-22, 1981, Gatlinburg, Tennessee, U.S.A., Extended Abstracts, p. 258.

61. H.C. Andersen, J. Chem. Phys. 72, 2384 (1980).

62. J.B. Boyce and T. Hayes, in Physics of Superionic Conductors (Ed. M.B. Salamon), Springer-Verlag, New York, 1979.

63. T. Hayes and J.B. Boyce, J. Phys. C: Solid St. Phys. 13, L731 (1980).

64. G.D. Mahan, in Superionic Conductors (Eds. G.D. Mahan and W.L. Roth), Plenum, New York, 1976, p. 115.

65. H.U. Beyeler, P. Bruesch, L. Pietronero, W.R. Schneider, S. Strassler and H.R. Zeller, in Physics of Superionic Conductors (Ed. M.B. Salamon), Springer-Verlag, New York, 1979, p. 77.

66. T. Geisel, in Physics of Superionic Conductors (Ed. M.B. Salamon), Springer-Verlag, New York, 1979, p. 201.

67. R. Zeyher, in Fast Ion Transport in Solids (Eds. P. Vashishta, J.N. Mundy and G.K. Shenoy), Elsevier North-Holland, 1979, p. 509.

68. L. Pietronero, S. Strassler and H.R. Zeller, in Fast Ion Transport in Solids (Eds. P. Vashishta, J.N. Mundy and G.K. Shenoy), Elsevier North-Holland, New York, 1979, p. 159.

SOME APPLICATIONS OF CONDITIONALLY CONVERGENT LATTICE SUMS

Edgar R. Smith[a] and John W. Perram[b]

[a]Department of Mathematics
University of Melbourne
Parkville, Vic. 3052
Australia

and

[b]Matematisk Institut
Odense Universitet
Campusvej 55
5230 Odense M
Denmark

ABSTRACT

We use some of the lattice sums developed in the previous lecture to study two systems. The first is the simple cubic lattice, with unit lattice spacing with a polarizable point of polarizability α at each lattice site. For this system we consider first a plane slab of the material in an external electric field. The response of the system to a small external field gives the dielectric constant of the system. We also calculate the response of the polarizable points close to the surface of the plane slab. Next we consider the response of a large sphere of the material to a charge or fixed dipole at the centre of the sphere and find an asymptotic expression for the polarization of points far from the "defect". These expansions also allow calculation of the dielectric constant. All methods give the same static dielectric constant

$$\varepsilon = (1+8\pi\alpha/3)(1-4\pi\alpha/3)$$

the standard Clausius-Mosotti result. The second system studied is a plane slab of ionic crystal responding to an applied electric field. We show how to calculate the distortion of the crystal produced by the field and thus obtain the dielectric constant of the crystal. Results for the dielectric constant for a CsCl structure are given and show how the dielectric constant depends on the repulsive potential between the ions.

I. THE POINT POLARIZABLE LATTICE

The first example we consider is a simple cubic lattice, of spacing 1 with a point of polarizability α at each lattice site. We shall see that the shape dependent terms in the lattice sums developed in the first lecture are crucial to obtaining a correct description of the properties of this lattice. This is the main point of studying such an artificially simple model. We begin by considering a piece P_N of the lattice L which has lattice vectors $\underset{\sim}{n} = (n_1,n_2,n_3)$. If an electric field $\underset{\sim}{e}(\underset{\sim}{n})$ is applied to the system, then the dipole moment $\underset{\sim}{\mu}(\underset{\sim}{m})$ at lattice site $\underset{\sim}{m}$ obeys the equation

$$\underset{\sim}{\mu}(\underset{\sim}{m}) = \alpha \underset{\sim}{e}(\underset{\sim}{m}) + \alpha \sum_{\underset{\sim}{n}\in P_N(L)} t(\underset{\sim}{m}-\underset{\sim}{n})\cdot\underset{\sim}{\mu}(\underset{\sim}{n}) \tag{1.1}$$

where

$$t(\underset{\sim}{m}) = \begin{cases} 0 & \text{if } \underset{\sim}{m} = \underset{\sim}{0} \\ +\nabla\nabla \left.\dfrac{1}{|\underset{\sim}{r}+\underset{\sim}{m}|}\right|_{\underset{\sim}{r}=\underset{\sim}{0}} & \text{if } \underset{\sim}{m} \neq \underset{\sim}{0} \end{cases} \tag{1.2}$$

$\alpha\nu\phi$

$$\nabla\nabla \left.\frac{1}{|\underset{\sim}{r}+\underset{\sim}{m}|}\right|_{\underset{\sim}{r}=\underset{\sim}{0}} = -\left[I - \frac{3\underset{\sim}{m}\underset{\sim}{m}}{|\underset{\sim}{m}|^2}\right] |\underset{\sim}{m}|^{-3} . \tag{1.3}$$

We shall study three cases:

(i) $P_N(L)$ a plane slab with large thickness, that is all lattice sites (n_1,n_2,n_3) with $-N \leq n_3 \leq N$, and $\underset{\sim}{e}(\underset{\sim}{m})$ an externally applied field of the form

$$\underset{\sim}{e}(\underset{\sim}{m}) = (0,0,1)E\, e^{2iKm_3} . \tag{1.4}$$

(ii) $P_N(L)$ a large sphere of radius N, that is, all lattice sites $\underset{\sim}{n}$ with $|\underset{\sim}{n}|^2 \leq N$, and $\underset{\sim}{e}(\underset{\sim}{m})$ the field set up by a charge Q placed on the central polarizable point so that

$$
\underset{\sim}{e}(\underset{\sim}{m}) = \begin{cases} -Q\nabla \left. \dfrac{1}{|\underset{\sim}{r}|} \right|_{\underset{\sim}{r}=\underset{\sim}{m}} = +Q \dfrac{\underset{\sim}{m}}{|\underset{\sim}{m}|^3} \, , & \underset{\sim}{m} \neq \underset{\sim}{0} \, , \\[2em] \underset{\sim}{0} \, , & \underset{\sim}{m} = \underset{\sim}{0} \, . \end{cases}
\tag{1.5}
$$

(iii) $P_N(L)$ the same as in (ii) but $\underset{\sim}{e}(\underset{\sim}{m})$ the field set up by a dipole $\underset{\sim}{\mu}$ placed on the central polarizable point, so that

$$
\underset{\sim}{e}(\underset{\sim}{m}) = \begin{cases} \nabla\nabla \left. \dfrac{1}{|\underset{\sim}{r}|} \right|_{\underset{\sim}{r}=\underset{\sim}{m}} \cdot \underset{\sim}{\mu} = -\left[I - \dfrac{3\underset{\sim}{m}\underset{\sim}{m}}{|\underset{\sim}{m}|^2} \right] |\underset{\sim}{m}|^{-3} \cdot \underset{\sim}{\mu} \, , & \underset{\sim}{m} \neq \underset{\sim}{0} \, , \\[2em] \underset{\sim}{0} \, , & \underset{\sim}{m} = \underset{\sim}{0} \, . \end{cases}
\tag{1.6}
$$

It is instructive to model these systems as continuum pieces of matter with dielectric constant ε, as these models will allow us to derive the dielectric constant of the lattice material from the solutions $\underset{\sim}{\mu}(\underset{\sim}{m})$ to eq. (1.1) [1].

(i) We have an electric field

$$
\underset{\sim}{e}(z) = (0,0,1)\, E\, e^{2iKz}
\tag{1.7}
$$

in the regions $z < -N$ and $< > N$ and the system is independent of x and y. Inside the slab, the displacement vector obeys

$$
\underset{\sim}{D}_i(z) = \underset{\sim}{E}_i(z) + 4\pi \underset{\sim}{P}(z) = \varepsilon(K)\, \underset{\sim}{E}_i(z)
\tag{1.8}
$$

where $\underset{\sim}{P}(z)$ is the polarization per unit volume and $\varepsilon(K)$ is a K-dependent dielectric constant. Continuity of the displacement vector gives

$$
\underset{\sim}{D}_i(z) = (0,0,1)\, E\, e^{2iKz}
\tag{1.9}
$$

so that

$$
\underset{\sim}{E}_i(z) = (0,0,1)\, \dfrac{E}{\varepsilon(K)}\, e^{2iKz} \, .
\tag{1.10}
$$

Thus

$$
\underset{\sim}{P}(z) = \dfrac{\varepsilon(K) - 1}{4\pi\varepsilon(K)}\, E\, e^{2iKz}\, (0,0,1) \, .
\tag{1.11}
$$

The connection between this macroscopic picture and the solution of the microscopic equation (1.1) is made by noting that the polarization density in a region around $\underset{\sim}{m}$ is simply $\underset{\sim}{\mu}(\underset{\sim}{m})$.

(ii) We have a sphere of continuous material of static dielectric constant ε of radius N, immersed in vacuum. The charge Q at the centre of the sphere sets up a potential $Q/|\underset{\sim}{r}|$ outside the sphere. Inside the sphere the potential is

$$\phi(\underset{\sim}{r}) = \frac{(\varepsilon-1)}{\varepsilon}\frac{Q}{N} + \frac{Q}{\varepsilon|\underset{\sim}{r}|} \tag{1.12}$$

so that the field inside the sphere is

$$\underset{\sim}{E}(\underset{\sim}{r}) = \frac{Q\underset{\sim}{r}}{\varepsilon|\underset{\sim}{r}|^3} \; . \tag{1.13}$$

The polarization density within the sphere is then

$$\underset{\sim}{P}(\underset{\sim}{r}) = \frac{\varepsilon-1}{4\pi\varepsilon} Q \frac{\underset{\sim}{r}}{|\underset{\sim}{r}|^3} \tag{1.14}$$

(iii) For this case the potential at $\underset{\sim}{r}$ outside the sphere is

$$\phi(\underset{\sim}{r}) = \frac{\underset{\sim}{\mu}\cdot\underset{\sim}{r}}{|\underset{\sim}{r}|^3} \tag{1.15}$$

while inside the sphere it is

$$\phi(\underset{\sim}{r}) = \left[\frac{2(\varepsilon-1)}{3\varepsilon N^3}|\underset{\sim}{r}| + \frac{(2+\varepsilon)}{3\varepsilon}|\underset{\sim}{r}|^{-2}\right]\frac{\underset{\sim}{\mu}\cdot\underset{\sim}{r}}{|\underset{\sim}{r}|} \; . \tag{1.16}$$

Thus inside the sphere the electric field is

$$\underset{\sim}{E}(\underset{\sim}{r}) = -\frac{2+\varepsilon}{3\varepsilon}\left[I - \frac{3\underset{\sim}{r}\underset{\sim}{r}}{|\underset{\sim}{r}|^2}\right]\frac{\underset{\sim}{\mu}}{|\underset{\sim}{r}|^3} - \frac{2(\varepsilon-1)}{3\varepsilon N^3}\underset{\sim}{\mu} \; . \tag{1.17}$$

In the limit $N \to \infty$, the polarization density is then

$$\underset{\sim}{P}(\underset{\sim}{r}) = -\frac{(\varepsilon+2)(\varepsilon-1)}{12\pi\varepsilon}\left[I - \frac{3\underset{\sim}{r}\underset{\sim}{r}}{|\underset{\sim}{r}|^2}\right]\frac{\underset{\sim}{\mu}}{|\underset{\sim}{r}|^3} \; . \tag{1.18}$$

We may now return to the solution of (1.1) for these cases. We define

$$\underset{\sim}{M}(\underset{\sim}{k}) = \sum_{\underset{\sim}{m}\in P_N(L)} \underset{\sim}{\mu}(\underset{\sim}{m})\, e^{2i\underset{\sim}{k}\cdot\underset{\sim}{m}}$$

$$\underset{\sim}{E}(\underset{\sim}{k}) = \sum_{\underset{\sim}{m}\in P_N(L)} \underset{\sim}{e}(\underset{\sim}{m})\, e^{2i\underset{\sim}{k}\cdot\underset{\sim}{m}}$$

and

$$T(\underset{\sim}{k}) = \sum_{\underset{\sim}{m}\in P_N(L)} t(\underset{\sim}{m})\, e^{2i\underset{\sim}{k}\cdot\underset{\sim}{m}} . \tag{1.19}$$

Using these Fourier transforms and the convolution theorem (1.25) of the first lecture, we obtain

$$[I - \alpha T(\underset{\sim}{k})]\cdot\underset{\sim}{M}(\underset{\sim}{k}) = \alpha\underset{\sim}{E}(\underset{\sim}{k}) . \tag{1.20}$$

This equation may be solved and the inverse transform ((1.23) of the first lecture) used to give

$$\underset{\sim}{\mu}(\underset{\sim}{m}) = \alpha\pi^{-3}\int_{V_{\pi/2}} d^3\underset{\sim}{k}[I - \alpha T(\underset{\sim}{k})]^{-1}\cdot\underset{\sim}{E}(\underset{\sim}{k})\, e^{-2i\underset{\sim}{k}\cdot\underset{\sim}{m}} . \tag{1.21}$$

Now we proceed to evaluate $\underset{\sim}{\mu}(\underset{\sim}{m})$ for the problems introduced above.

(i) $P_N(L)$ a plane slab.

For this case

$$\underset{\sim}{E}(\underset{\sim}{k}) = \sum_{\underset{\sim}{m}\in P_N(L)} (0,0,1)E\, e^{2iKm_3}\, e^{2ik\cdot\underset{\sim}{m}} . \tag{1.22}$$

The sums over m_1 and m_2 are unrestricted and so give delta functions [2]. The sum over m_3 is a geometric series, so that

$$\sum_{\underset{\sim}{m}\in P_N(L)} e^{2iKm_3+2i\underset{\sim}{k}\cdot\underset{\sim}{m}} = \pi^2\delta(k_1)\delta(k_2)\sum_{m=-N}^{N} e^{2i(K+k_3)m}$$

$$= \pi^2\delta(k_1)\delta(k_2)\frac{\sin(2N+1)(K+k_3)}{\sin(K+k_3)} . \tag{1.23}$$

and so

$$\underset{\sim}{E}(\underset{\sim}{k}) = \pi^2(0,0,1)E\delta(k_1)\delta(k_2)\frac{\sin(2N+1)(K+k_3)}{\sin(K+k_3)} . \tag{1.24}$$

This may be substituted into eq. (1.21) to give $\underset{\sim}{\mu}(\underset{\sim}{m})$. For N large, and $|m_3| \ll N$, the function $\sin(2N+1)x/\sin x$ acts as $\pi\delta(x)$ so that in the centre of the slab,

$$\underset{\sim}{\mu}(\underset{\sim}{m}) = \alpha[I - \alpha T(0,0,-K)]^{-1}(0,0,1)E\, e^{2iKm_3} . \tag{1.25}$$

For $K \neq 0$, $T(0,0,-K)$ is given by (4.11, 13, 17) of the first lecture:

$$T(0,0,-K) = \frac{4y^3}{3\sqrt{\pi}} I + \sum_{\substack{n \in L \\ n \neq 0}} \left\{ \left(\frac{3nn}{n^2} - I \right) \left(\frac{\operatorname{erfc}(y|n|)}{|n|^3} + \frac{2y}{\sqrt{\pi} n^2} e^{-y^2 n^2} \right) + \frac{4y^3}{\sqrt{\pi}} \frac{nn}{n^2} e^{-y^2 n^2} \right\} e^{-2iKn_3}$$
$$- 4\pi \sum_{\substack{m \in R \\ m \neq 0}} \frac{[\pi m + (0,0,K)][\pi m + (0,0,K)]}{[\pi m + (0,0,K)]^2} \times e^{-[\pi m + (0,0,K)]^2 / y^2}$$
$$- 4\pi S e^{-K^2/y^2} \tag{1.26}$$

where S is the matrix

$$S = \begin{pmatrix} 0 & 0 & 0 \\ 0 & 0 & 0 \\ 0 & 0 & 1 \end{pmatrix} . \tag{1.27}$$

This expression may be evaluated numerically and inserted into eq. (1.25). Comparison with eq. (1.11) then gives $\varepsilon(K)$. In figs 1 we plot $\varepsilon(K)$ as a function of α for several values of K and $\varepsilon(K)$ as a function of K for several values of α. The value of α, $\alpha_c(K)$, for which $\varepsilon(K)$ diverges, the well-known polarization catastrophe, is plotted in figure 2.

For $K = 0$, we have

$$\mu(m) = \alpha[I - \alpha T(0,0,0)]^{-1} (0,0,E) , \tag{1.28}$$

a constant, except near the surfaces of the slab. We may evaluate $T(0)$ analytically for a plane slab. We have, for a spherical $P_N(L)$ [3],

$$T(0) = \sum_{\substack{n \in L \\ n \neq 0}} \nabla\nabla \frac{1}{|r|} \bigg|_{r=n} = 0 , \tag{1.29}$$

this result following from the symmetry of the summand for a spherical $P_N(L)$. Also, for a spherical $P_N(L)$,

$$T(0) = \nabla\nabla\psi_E(r,L) \bigg|_{r=0} - \frac{4\pi I}{3} . \tag{1.30}$$

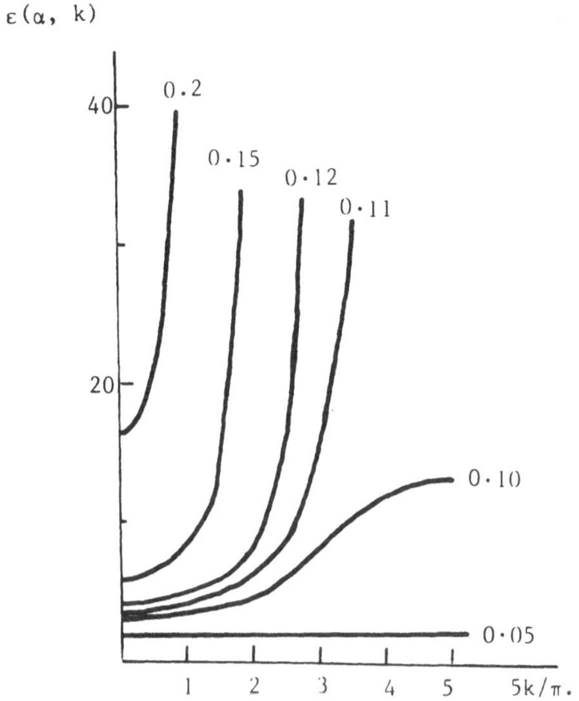

Fig.1a. Plot of ε(α,k) for various values of α for point polarizable lattice.

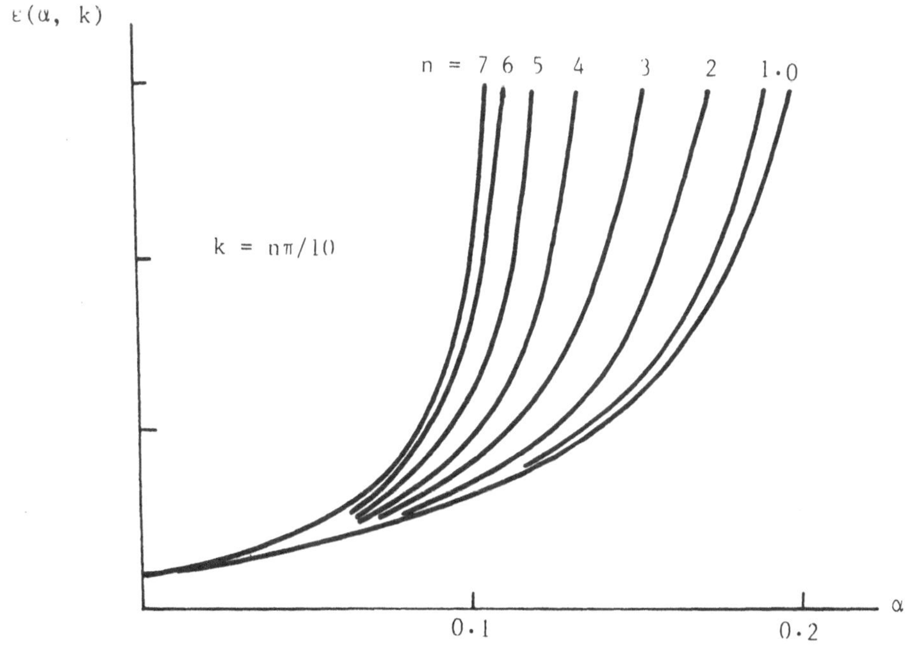

Fig.1b. Plot of ε(α,k) for various values of k for point polarizable lattice.

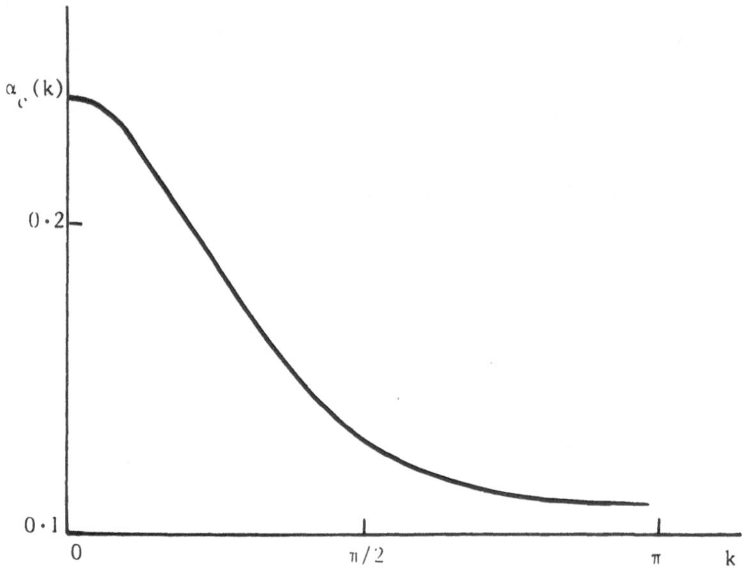

Fig. 2. Plot of $\alpha_c(k)$ on $0 < k < \pi$.

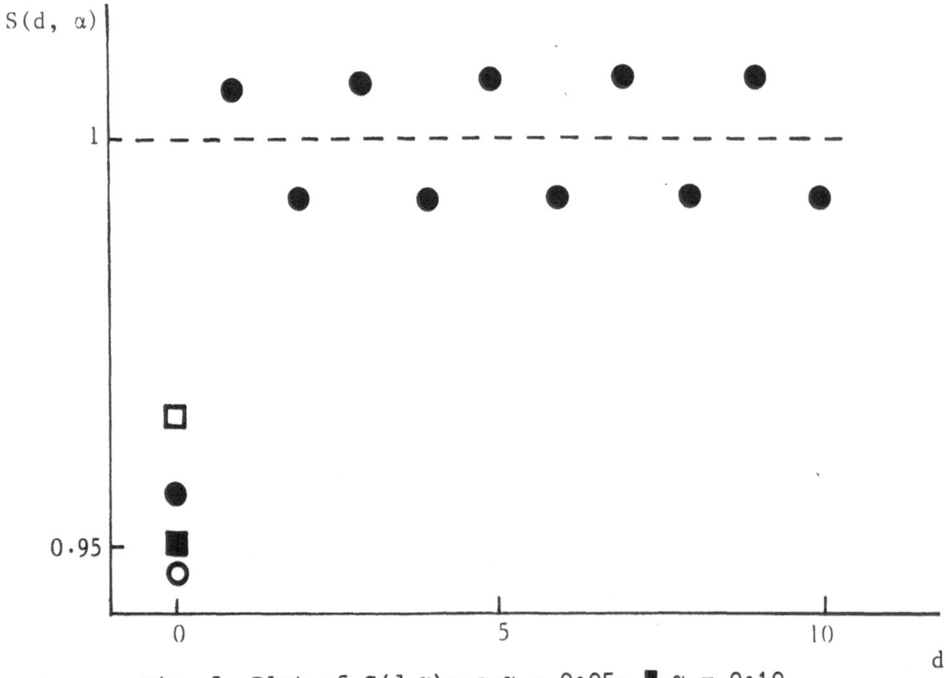

Fig. 3. Plot of $S(d,\alpha)$: ●, $\alpha = 0{\cdot}05$; ■, $\alpha = 0{\cdot}10$, ▲, $\alpha = 0{\cdot}15$; o, $\alpha = 0{\cdot}20$.

Thus

$$\nabla\nabla\psi_E(\underset{\sim}{r},L)\Big|_{\underset{\sim}{r}=\underset{\sim}{0}} = \frac{4\pi I}{3} \quad . \tag{1.31}$$

For a planar slab $P_N(L)$,

$$\begin{aligned} T(\underset{\sim}{0}) &= \nabla\nabla\psi_E(\underset{\sim}{r},L)\Big|_{\underset{\sim}{r}=\underset{\sim}{0}} - 4\pi S \\ &= \frac{4\pi}{3} I - 4\pi S \ . \end{aligned} \tag{1.32}$$

Thus

$$[I - \alpha T(\underset{\sim}{0})]^{-1} = \begin{bmatrix} 1 - \frac{4\pi\alpha}{3} & 0 & 0 \\ 0 & 1 - \frac{4\pi\alpha}{3} & 0 \\ 0 & 0 & 1 + \frac{8\pi\alpha}{3} \end{bmatrix}^{-1} \quad . \tag{1.33}$$

Thus for $\underset{\sim}{m}$ not close to a surface of the planar slab, with K = 0,

$$\underset{\sim}{\mu}(\underset{\sim}{m}) = \frac{\alpha}{1 + 8\pi\alpha/3} E(0,0,1) \ . \tag{1.34}$$

If we compare this with eq. (1.11) we obtain

$$\varepsilon(0) = (1 + 8\pi\alpha/3)(1 - 4\pi\alpha/3) \ , \tag{1.35}$$

the usual Clausius-Mosotti expression.

It is interesting to ask what happens when we leave out the shape dependence, or use the shape dependence of a sphere. If we leave out the shape dependence,

$$[I - \alpha T(\underset{\sim}{0})]^{-1} = \frac{1}{1 - 4\pi\alpha/3} I \ , \tag{1.36}$$

$$\underset{\sim}{\mu}(\underset{\sim}{m}) = \frac{\alpha E}{1 - 4\pi\alpha/3} (0,0,1) \tag{1.37}$$

and

$$\varepsilon(0) = \frac{1 - 4\pi\alpha/3}{1 - 16\pi\alpha/3} \quad . \tag{1.38}$$

Curiously, this result is correct to order α^1 but higher order powers of α are incorrect. If we use the shape dependence of a sphere,

$$[I - \alpha T(\underset{\sim}{0})]^{-1} = I \tag{1.39}$$

then

$$\varepsilon(0) = (1 - 4\pi\alpha)^{-1} \tag{1.40}$$

which is also correct to order α^1 with incorrect higher powers of α.

We may now return to eq. (1.21) using eq. (1.24) for the case when m is close to the $m_3 = -N$ surface of the slab. We consider $K = \underset{\sim}{0}$ and $\underset{\sim}{m} = (0,0,-N+d)$. Note that for this case

$$\frac{\sin(2N+1)k_3}{\sin k_3} e^{2ik_3(N-d)} = \frac{1}{2} \cos 2k_3 d \left[\frac{\sin(4N+1)k_3}{\sin k_3} + 1\right] - \frac{\sin 2k_3 d}{2 \sin k_3}[\cos k_3 - \cos(4N+1)k_3]$$
$$+ \text{ imaginary terms which are odd in } k_3. \tag{1.41}$$

The rest of the integrand in eq. (1.21) is even in k_3 so that the imaginary terms in eq. (1.41) are absent. As $N \to \infty$, the term with $\sin(4N+1)k_3$ behaves as $\pi\delta(k_3)$ while the $\cos(4N+1)k_3$ term gives zero contribution by the Riemann-Lebesgue lemma. Thus we obtain, for $\underset{\sim}{m} = (0,0,-N+d)$

$$\underset{\sim}{\mu}(\underset{\sim}{m}) = \frac{1}{2}\underset{\sim}{\mu}(\underset{\sim}{0}) + \frac{\alpha E}{2\pi}\int_{-\pi/2}^{\pi/2} dk[I - \alpha T(0,0,k)]^{-1}_{3,3} \times \frac{\sin(1-2d)k_3}{\sin k_3}(0,0,1) \tag{1.42}$$

where

$$\underset{\sim}{\mu}(\underset{\sim}{0}) = \frac{\alpha E}{1 + 8\pi\alpha/3}(0,0,1) \tag{1.43}$$

is the polarization deep within the slab. We may write eq. (1.42) in the form

$$\underset{\sim}{\mu}(\underset{\sim}{m}) = \frac{\alpha E}{1 + 8\pi\alpha/3} S(d,\alpha)(0,0,1) \tag{1.44}$$

and calculate $S(d,\alpha)$ numerically. It is presented in figure 3. An important feature of $S(d,\alpha)$ is that except for $d = 0$, it is within 0•01 of its bulk value 1, and at $d = 0$ it is within 0•05 of its bulk value. This suggests that the long-range dipole-dipole interaction acts as a very short-ranged effective interaction plus a long-ranged part which seems rather insensitive to structural details.

(ii) $P_N(L)$ a large sphere and

$$\underset{\sim}{e}(\underset{\sim}{m}) = \begin{cases} -Q\underset{\sim}{m}/|\underset{\sim}{m}|^3 , & \underset{\sim}{m} \neq \underset{\sim}{0} \\ \underset{\sim}{0} , & \underset{\sim}{m} = \underset{\sim}{0} \end{cases} , \tag{1.45}$$

the field due to a point charge Q on the central polarizable point. For this case we consider only $\underset{\sim}{\mu}(\underset{\sim}{m})$ for large $|\underset{\sim}{m}|$ and develop an asymptotic expansion of it in inverse powers of $|\underset{\sim}{m}|$. For this case

$$\begin{aligned} \underset{\sim}{E}(\underset{\sim}{k}) &= -Q\nabla \sum_{\underset{\sim}{m}\in P_N(L)} |\underset{\sim}{r}+\underset{\sim}{m}|^{-1}\Big|_{\underset{\sim}{r}=\underset{\sim}{0}} e^{2i\underset{\sim}{k}\cdot\underset{\sim}{m}} \\ &= -Q\nabla\, \theta(\underset{\sim}{r},\underset{\sim}{k};P,L,N)\Big|_{\underset{\sim}{r}=\underset{\sim}{0}} \end{aligned} \tag{1.46}$$

where $\theta(\underset{\sim}{r},\underset{\sim}{k};P,L,N)$ is given by eq. (4.3) of the first lecture. If $|\underset{\sim}{m}|$ is large, then we may change the variable of integration $\underset{\sim}{k}$ in eq. (1.21) to $\underset{\sim}{w} = |\underset{\sim}{m}|\underset{\sim}{k}$ and obtain

$$\underset{\sim}{\mu}(\underset{\sim}{m}) = \frac{\alpha}{\pi^3|\underset{\sim}{m}|^3} \int_{V_{|\underset{\sim}{m}|\pi/2}} d^3\underset{\sim}{w}[I - \alpha T(\underset{\sim}{w}/|\underset{\sim}{m}|)]^{-1}\, \underset{\sim}{E}\left(\frac{\underset{\sim}{w}}{|\underset{\sim}{m}|}\right) e^{-2i\underset{\sim}{w}\cdot\hat{\underset{\sim}{m}}} . \tag{1.47}$$

We may now make an expansion of the functions of w/ m in the integrand of eq. (1.47) and keep only the leading terms. For a spherical $P_N(L)$ for small $\underset{\sim}{k}$, .

$$T(\underset{\sim}{k}) = \frac{4\pi}{3} I - 4\pi \frac{\underset{\sim}{k}\underset{\sim}{k}}{\underset{\sim}{k}^2} + O(\underset{\sim}{k}^2) \tag{1.48}$$

so that to leading order,

$$[I - \alpha T(\underset{\sim}{k})]^{-1} = \left[\left(1 - \frac{4\pi\alpha}{3}\right) I + 4\pi\alpha \frac{\underset{\sim}{k}\underset{\sim}{k}}{\underset{\sim}{k}^2}\right]^{-1} . \tag{1.49}$$

If we note that $(\underset{\sim}{k}\underset{\sim}{k}/\underset{\sim}{k}^2)^2 = \underset{\sim}{k}\underset{\sim}{k}/\underset{\sim}{k}^2$ then the result

$$[I - \alpha T(\underset{\sim}{k})]^{-1} = \frac{1}{1 - 4\pi\alpha/3} I - \frac{4}{(1+8\pi\alpha/3)(1-4\pi\alpha/3)} \frac{\underset{\sim}{k}\underset{\sim}{k}}{\underset{\sim}{k}^2} + O(\underset{\sim}{k}^2)$$

(1,50)

may be checked by multiplying the left side by $I - \alpha T(\underset{\sim}{k})$. If we note that eq. (4.3) of the first lecture shows $\theta(\underset{\sim}{r},\underset{\sim}{k};P,L,N)$ to be even in $\underset{\sim}{r}$ for $\underset{\sim}{k} = \underset{\sim}{0}$ we obtain

$$\underset{\sim}{E}(\underset{\sim}{k}) = 2\pi i Q \frac{\underset{\sim}{k}}{k^2} + O(\underset{\sim}{k}) \ . \tag{1.51}$$

We may substitute this and eq. (1.50) into eq. (1.47) and obtain for large $|\underset{\sim}{m}|$,

$$\underset{\sim}{\mu}(\underset{\sim}{m}) = \frac{2\alpha Q}{1 + 8\pi\alpha/3} \frac{1}{\pi^2} \int_{V_{\pi/2}} \frac{\underset{\sim}{k}}{k^2} \sin 2\underset{\sim}{k}\cdot\underset{\sim}{m}\, d^3\underset{\sim}{k} + O(|\underset{\sim}{m}|^{-4}). \tag{1.52}$$

This integral may be evaluated fairly easily and we obtain the asymptotic representation

$$\underset{\sim}{\mu}(\underset{\sim}{m}) = \frac{\alpha Q}{1 + 8\pi\alpha/3} \frac{\underset{\sim}{m}}{|\underset{\sim}{m}|^3} + O(|\underset{\sim}{m}|^{-4}) \ . \tag{1.53}$$

If we compare this polarization with that for a sphere of continuum dielectric with a charge Q at its centre we obtain

$$\varepsilon(0) = (1 + 8\pi\alpha/3)(1 - 4\pi\alpha/3) \tag{1.54}$$

which is the Clausius-Mosotti result again.

(iii) $P_N(L)$ a large sphere and

$$\underset{\sim}{e}(\underset{\sim}{m}) = \begin{cases} -\, I - \dfrac{3\underset{\sim}{m}\underset{\sim}{m}}{|\underset{\sim}{m}|^2} \dfrac{\underset{\sim}{\mu}}{|\underset{\sim}{m}|^3}\ , & \underset{\sim}{m} \neq \underset{\sim}{0} \\ \underset{\sim}{0}\ , & \underset{\sim}{m} = \underset{\sim}{0} \end{cases} \ , \tag{1.55}$$

the field due to a dipole $\underset{\sim}{\mu}$ at the centre of the sphere. For this system

$$\underset{\sim}{E}(\underset{\sim}{k}) = T(\underset{\sim}{k})\cdot\underset{\sim}{\mu} \tag{1.56}$$

so that eq. (1.21) may be written using

$$\alpha\underset{\sim}{E}(\underset{\sim}{k}) = \underset{\sim}{\mu} - [I - \alpha T(\underset{\sim}{k})]\cdot\underset{\sim}{\mu} \tag{1.57}$$

as

$$\underset{\sim}{\mu}(\underset{\sim}{m}) = \pi^{-3} \int_{V_{\pi/2}} d^3\underset{\sim}{k}[I - \alpha T(\underset{\sim}{k})]^{-1} e^{-2i\underset{\sim}{k}\cdot\underset{\sim}{m}} d^3\underset{\sim}{k} + \underset{\sim}{\mu}\delta_{\underset{\sim}{m},\underset{\sim}{0}} . \tag{1.58}$$

Using eq. (1.50) we obtain for large $|\underset{\sim}{m}|$,

$$\underset{\sim}{\mu}(\underset{\sim}{m}) = -\frac{\alpha}{(1+8\pi\alpha/3)(1-4\pi\alpha/3)}\frac{4}{\pi^2}\int_{V_{\pi/2}}\frac{\underset{\sim}{k}\underset{\sim}{k}\cdot\underset{\sim}{\mu}}{\underset{\sim}{k}^2}e^{-2i\underset{\sim}{k}\cdot\underset{\sim}{m}}\,d^3\underset{\sim}{k}$$

$$+\ O(|\underset{\sim}{m}|^{-5})\ . \tag{1.59}$$

The integral in eq. (1.59) may be evaluated for large $|\underset{\sim}{m}|$ by writing

$$\nabla\nabla e^{-2i\underset{\sim}{k}\cdot\underset{\sim}{r}} = -4\ \underset{\sim}{k}\underset{\sim}{k}\ e^{2i\underset{\sim}{k}\cdot\underset{\sim}{r}} \tag{1.60}$$

and using the identity

$$\underset{\sim}{k}^{-2} = \int_0^\infty dt\ e^{-t\underset{\sim}{k}^2}\ . \tag{1.61}$$

We obtain [3]

$$\underset{\sim}{\mu}(\underset{\sim}{m}) = -\frac{\alpha}{(1+8\pi\alpha/3)(1-4\pi\alpha/3)}\left(I - \frac{3\underset{\sim}{m}\underset{\sim}{m}}{|\underset{\sim}{m}|^2}\right)\frac{\underset{\sim}{\mu}}{|\underset{\sim}{m}|^3} + O(|\underset{\sim}{m}|^{-5}). \tag{1.62}$$

This expression may be compared with its continuum analogue eq. (1.18). We obtain

$$\frac{(\varepsilon+2)(\varepsilon-1)}{\varepsilon} = \frac{9x}{(1+2x)(1-x)} \tag{1.63}$$

with $x = 4\pi\alpha/3$. The resulting quadratic equation for ε is

$$\varepsilon^2 = \frac{(1-8x-2x^2)}{1+x-2x^2}\ \varepsilon - 2 = 0\ . \tag{1.64}$$

To obtain a positive dielectric constant we must take the larger root

$$\varepsilon = -\frac{(1-8x-2x^2)}{2(1+2x)(1-x)} + \sqrt{\frac{(1-8x-2x^2)^2}{4(1+2x)^2(1-x)^2} + 2}\ . \tag{1.65}$$

Some rather tedious simplification gives

$$\varepsilon = \frac{1+8\pi\alpha/3}{1-4\pi\alpha/3} \tag{1.66}$$

again.

In studying this simple model example at some length we have shown the response of a simple model crystal to various simple distorting fields. We have seen that the shape dependent terms in the lattice sums which are involved determine the answer obtained in part. Further, we have shown that by correctly accounting for the shape dependent part of the lattice sums, a reliable dielectric constant for the lattice may be calculated. The result is obtained from the response of a plane slab to an external field, or the response of a sphere to a point charge or dipole embedded in it. The same result may also be obtained by considering the response of a sphere to a constant external field [4]. To obtain this consticency it was necessary to take proper account of the shape dependence of the lattice sums and of the correct solutions of Laplace's equation for the continuum model results which gave the connection between polarization density and dielectric constant for the various shapes and distorting fields considered.

II. THE SIMPLE IONIC LATTICE

In this section we return to the equations for the distortion of a simple ionic lattice (which may include shell model ions to take account of polarizability effects) when some extra potential is added to the system. In the first lecture we found that the shift $\underset{\sim}{s}_k(\underset{\sim}{m})$ of an ion of species k in the unit cell at $\underset{\sim}{m}$ in a piece $P_N(L)$ of crystal, subject to extra potentials $\psi_k(\underset{\sim}{r})$ satisfies the equation

$$\nabla\psi_k(\underset{\sim}{r}_k+\underset{\sim}{m}+\underset{\sim}{s}_k(\underset{\sim}{m})) + \sum^*_{\underset{\sim}{n}\in P_N(L)} \sum_{j=1}^{H} \nabla\Phi_{kj}{}^T(\underset{\sim}{r}_{kj}+\underset{\sim}{m}-\underset{\sim}{n}+\underset{\sim}{s}_k(\underset{\sim}{m})-\underset{\sim}{s}_j(\underset{\sim}{n}))$$
$$= \underset{\sim}{0} \,. \tag{2.1}$$

The potential $\phi_{kj}{}^T(\underset{\sim}{r})$ contains both the short-ranged and Coulomb interaction between ions of type k and j. For the case we consider here, which is a plane slab with a constant external field $\underset{\sim}{E}$ applied,

$$\psi_k(\underset{\sim}{r}) = -q_k\,\underset{\sim}{E}\cdot\underset{\sim}{r} \,, \tag{2.2}$$

so that $\psi_k(\underset{\sim}{r})$ is precisely linear in $\underset{\sim}{r}$. We may make the external field as small as we like, so that the approximation

$$\nabla\phi_{kj}{}^T(\underset{\sim}{r}_{kj}+\underset{\sim}{m}-\underset{\sim}{n}+\underset{\sim}{s}_k(\underset{\sim}{m})-\underset{\sim}{s}_j(\underset{\sim}{n})) = \nabla\phi_{kj}{}^T(\underset{\sim}{r}_{kj}+\underset{\sim}{m}-\underset{\sim}{n})$$
$$+ (\underset{\sim}{s}_k(\underset{\sim}{m})-\underset{\sim}{s}_j(\underset{\sim}{n}))\cdot\nabla\nabla\phi_{kj}{}^T(\underset{\sim}{r}_{kj}+\underset{\sim}{m}-\underset{\sim}{n}) \tag{2.3}$$

is correct. In this linearized approximation, eq. (2.1) becomes

$$\sum_{\underset{\sim}{n}\in P_N(L)}^{*} \sum_{j=1}^{H} [\underset{\sim}{s}_k(\underset{\sim}{m})-\underset{\sim}{s}_j(\underset{\sim}{n})]\left\{\nabla\nabla\phi_{kj}(\underset{\sim}{r}_{kj}+\underset{\sim}{m}-\underset{\sim}{n}) + q_k q_j \nabla\nabla|\underset{\sim}{r}_{kj}+\underset{\sim}{m}-\underset{\sim}{n}|^{-1}\right\} = q_k \underset{\sim}{E} . \tag{2.4}$$

The first term on the right-hand side of eq. (2.3) does not contribute because the undistorted lattice must be in equilibrium. The solution to eq. (2.4) will give the excess polarization per unit cell

$$\underset{\sim}{M}_E(\underset{\sim}{n}) = \sum_{i=1}^{H} q_i \underset{\sim}{s}_i(\underset{\sim}{n}) \tag{2.5}$$

caused by the electric field. From our experience with the point polarizable lattice we may assume that $\underset{\sim}{s}_i(\underset{\sim}{n})$ is independent of $\underset{\sim}{n}$ except in the two surface layers. We must also assume that the plane slab of crystal does have properly adjusted surface structure if the unit cell has a net dipole moment with component normal to the surfaces. The charges associated with these adjusted surface layers must develop a dipole moment which cancels that of the undistorted unit cells. Thus the polarization density in the interior of the crystal is

$$\underset{\sim}{M}_E = \sum_{i=1}^{H} q_i \underset{\sim}{s}_i \tag{2.6}$$

where the $\underset{\sim}{s}_i$ satisfy the equation

$$\sum_{j=1}^{H} [\Phi_{kj}(\underset{\sim}{0})+T_{kj}(\underset{\sim}{0})]\cdot(\underset{\sim}{s}_k-\underset{\sim}{s}_j) = q_k \underset{\sim}{E} , \tag{2.7}$$

where $\Phi_{kj}(\underset{\sim}{k})$ and $T_{kj}(\underset{\sim}{k})$ are defined in eqs. (1.27d and e) of the first lecture. We may write these equations in the form

$$\sum_{j=1}^{H} R_{kj} \underset{\sim}{s}_j = q_k \underset{\sim}{E} \tag{2.8}$$

where

$$R_{kj} = \sum_{i=1}^{H} (\Phi_{ki}(\underset{\sim}{0}) + T_{ki}(\underset{\sim}{0}))\delta_{kj} - \Phi_{kj}(\underset{\sim}{0}) - T_{kj}(\underset{\sim}{0}) . \tag{2.9}$$

Note that

$$\sum_{j=1}^{H} R_{kj} = 0 \tag{2.10}$$

so that eq. (2.9) does not have a solution. This is because the energy of the crystal is unaffected by movement of the whole crystal slab in the field. Accordingly we may choose $\underset{\sim}{s}_H = \underset{\sim}{0}$ and then solve

$$\sum_{j=1}^{H-1} R_{kj}\, \underset{\sim}{s}_j = q_k \underset{\sim}{E} \tag{2.11}$$

where R_{kj} is an $(H-1)\times(H-1)$ matrix with 3×3 matrix entries given by eq. (2.9). Thus

$$\underset{\sim}{s}_k = \begin{cases} \left(\sum_{j=1}^{H-1} R^{-1}{}_{kj}\, q_j\right) \underset{\sim}{E}\,, & k = 1,\dots,H-1 \\ \underset{\sim}{0}\,, & k \equiv H \end{cases} \tag{2.12}$$

and so

$$\underset{\sim}{M}_{\underset{\sim}{E}} = \sum_{k=1}^{H-1} \sum_{j=1}^{H-1} R^{-1}{}_{kj}\, q_j q_k \underset{\sim}{E}\,. \tag{2.13}$$

The dielectric constant may be obtained from the equations for the polarization of a plane slab of material (eqs. 1.8, 9, 10, 11). They give

$$\underset{\sim}{M}_{\underset{\sim}{E}} = \frac{\varepsilon-1}{4\pi\varepsilon} \underset{\sim}{E} \tag{2.14}$$

whence

$$\varepsilon = \left[1 - 4\pi \sum_{k=1}^{H-1} \sum_{j=1}^{H-1} R^{-1}{}_{kj}\, q_k q_j\right]^{-1}. \tag{2.15}$$

We now require the elements of R. The short-ranged interaction between ions i and j gives

$$\Phi_{ij}(\underset{\sim}{0}) = \sum_{\underset{\sim}{m}\in L} \nabla\nabla\phi_{ij}(\underset{\sim}{r}_{ij}+\underset{\sim}{m})\,. \tag{2.16}$$

The lattice sum is an unrestricted one over $\underset{\sim}{m} \in L$ because $\phi_{ij}(\underset{\sim}{r})$ is short-ranged and the sum is thus absolutely convergent. On the other hand, for a plane slab,

$$T_{kj}(\underset{\sim}{0}) = q_j q_j \sum^*_{\underset{\sim}{m} \in P_N(L)} \nabla\nabla |\underset{\sim}{r}_{kj} + \underset{\sim}{m}|^{-1}$$

$$= q_k q_j \left\{ \nabla\nabla\psi_E(\underset{\sim}{r}_{kj}, L) - \frac{4\pi}{|L|} S \right\} \tag{2.17}$$

where the matrix S is given in eq. (1.27). We may now compute ε from the response of a plane slab of crystal to an external field given the interparticle potentials.

We now consider a simple model crystal which has a simple cubic unit cell with the CsCl structure. That is, positive charges -1 at $\ell(n_1 + \frac{1}{4}, n_2 + \frac{1}{4}, n_3 + \frac{1}{4})$ and negative charges -1 at $\ell(n_1 - \frac{1}{4}, n_2 - \frac{1}{4}, n_3 - \frac{1}{4})$. Here, ℓ is the lattice spacing. The repulsive short-ranged interactions are all assumed equal, and to be

$$\phi_{kj}(\underset{\sim}{r}) = \frac{A\, e^{-\mu|\underset{\sim}{r}|}}{|\underset{\sim}{r}|} \tag{2.18}$$

It is interesting to study this model to see how the parameters A and μ affect the dielectric constant. In figure 4 we plot

$$\phi(r) = \frac{A\, e^{-\mu r}}{r} - \frac{1}{r}$$

for several values of A and μ. In figures 5 and 6 we plot the dielectric constants as a function of A and μ. Notice from figure 4 that as μ increases, the lattice becomes more tightly bound, or rigid and that the same happens as A decreases. As the lattice becomes more rigid, the dielectric constant decreases. For large enough μ or small enough A, the dielecrric constant will diverge. It would appear that this is a form of polarization catastrophe and possibly the ions find that the CsCl structure is only weakly metastable. Calculating the right-hand side of eq. (2.15) is effectively calculating the linear response of the lattice to a unit clectric field and it seems likely that the divergence is caused in part by the unit electric field attempting to change the structure of the ionic lattice.

Finally we show in figure 7 calculations of ε using an incorrect value of the lattice spacing. When the lattice is fairly rigid, errors in the lattice spacing cause only small errors in ε. However, as the lattice becomes more floppy, ε can be badly affected by quite small errors in the lattice spacing.

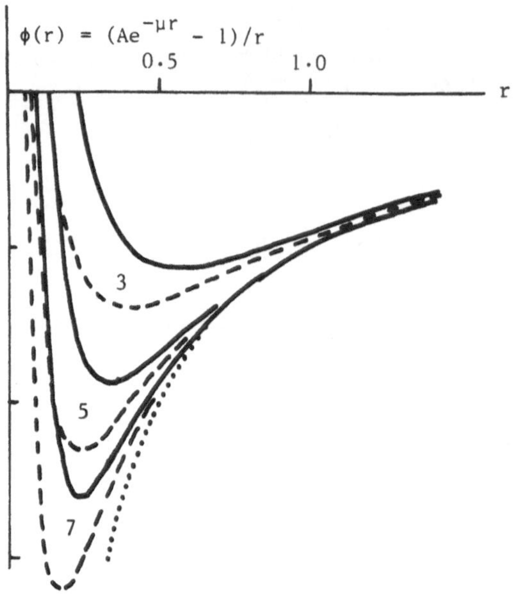

Fig. 4. Plot of pair potential for model ionic lattice
-A = 1•5,..., A = 2•0. Labels refer to values of μ.

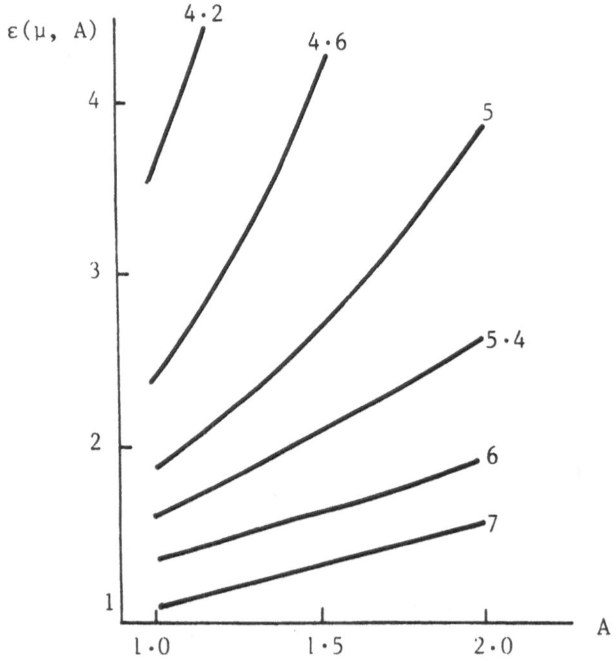

Fig. 5. Plot of ε(μ,A) for model ionic lattice. Labels refer to values of μ.

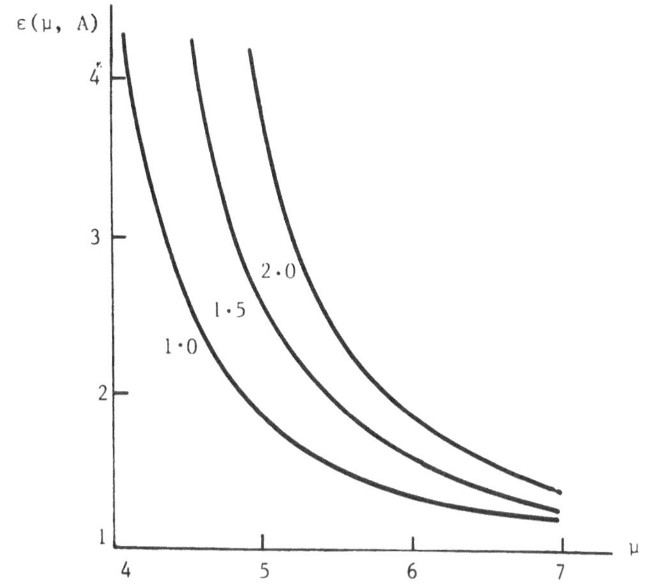

Fig.6. Plot of ε(μ,A) for model ionic lattice. Labels refer to values of A.

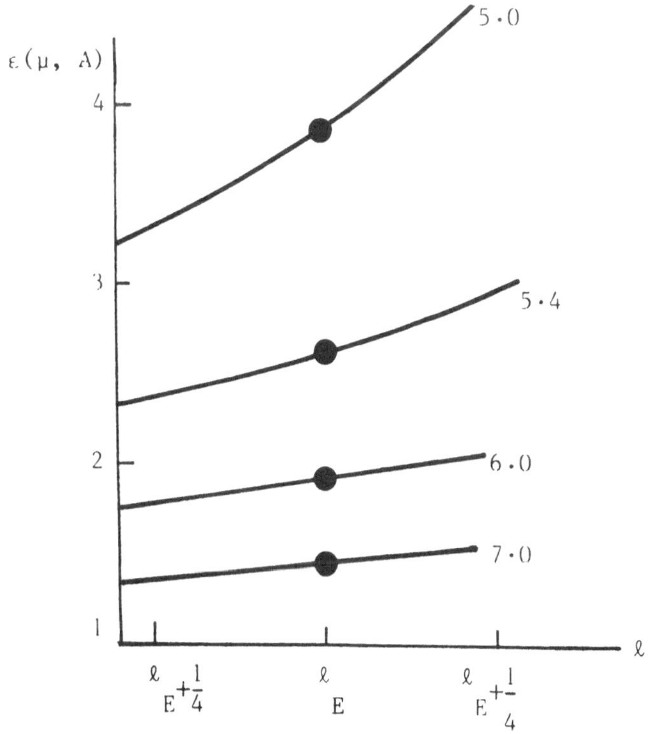

Fig.7. Plot of ε(μ,A) for model ionic lattice with incorrect lattice spacing ℓ,A=2. Labels refer to values of μ.

III. CONCLUSION

The point of the calculations described here is to show how the lattice sum equations for the distortion of crystals by external fields and defects are affected by the conditional convergence of the sums over Coulomb and dipole potentials. That the polarization density in a piece of dielectric should be dependent on shape is well known, but it is important to understand firstly that these effects are contained within the lattice sum equations and don't need to be added on separately and secondly that there are techniques which allow these effects to be calculated.

REFERENCES

1. Böttcher, C.J.F., "Theory of Electric Polarization, vol. 1", Elsevier, Masterdam (1973). See chapters 3,4.
2. Lighthill, M.J., "Introduction to Fourier Analysis and Generalized Functions", Cambridge (1958).
3. Smith, E.R., J. Phys. A. (Math. and Gen.) 13 (1980), L107-L110.
4. Böttcher, C.J.F., ibid., chapter 6.

BOUNDARY CONDITIONS IN THE SIMULATION OF IONIC SYSTEMS

S.W. de Leeuw

Theoretical Physics
1 Keble Road
Oxford

ABSTRACT

The Hamiltonian of a periodic array of charges contains terms depending on the shape and the surroundings of the system. The effect of these terms on the properties of the system is considered and some relations are derived for the long wavelength behaviour of fluctuations in the system. A brief discussion of the dynamical properties of the system under different boundary conditions is given.

1. INTRODUCTION

Computer simulation techniques are by now standard tools for the study of a wide variety of phenomena in solid and liquid state physics. It has recently become clear however that special care is required when applying these techniques to ionic and polar systems. The long range nature of the interactions in such systems implies that each particle interacts not only with a limited number of neighbouring particles, as in a neutral system, but with all other particles. It is well known that this leads to a strong collective behaviour in long wavelength phenomena. For periodic systems, which are normally used in computer simulations, the long range interactions imply a number of special features, which are not always recognized. For example the expression for the energy of such a system contains terms depending on the shape of the system and the nature of its surroundings [1-4]. For polar systems it has been shown that these terms can have a drastic effect on the long wavelength behaviour of fluctuations [1,5,6]. In particular it was shown that, for a spherical system surrounded by a medium of dielectric constant ε', the fluctuations in the dipole moment

M of the simulation cell depend strongly on ε'. This dependence on ε' nonetheless leads to a dielectric constant ε of the system independent of ε'.

It is of course tempting to bypass the difficulties arising from the long range interactions by introducing an arbitary cut-off radius beyond which the interaction between pairs is set equal to zero. Such a procedure is adapted in simulations which use the minimum image convention [7]. Such simulations can be useful for the study of a number of thermodynamic quantities and some structural properties. However, it is not possible in these computer experiments to make the connection with macroscopic electrostatics. Consequently one cannot study a large number of properties, in particular those properties which distinguish such systems from neutral systems using minimum image simulations.

In this chapter we shall consider the effects of boundary conditions on the thermodynamic, structural and dynamic properties of ionic systems. We shall confine ourselves to spherical systems surrounded by a medium of dielectric constant ε', and discuss the dependence of some physical quantities of the system on ε'.

2. THE METHOD OF MOLECULAR DYNAMICS

The techniques of computer simulation were first applied to ionic systems by Brush et al. [8], who studied the one component plasma, and Woodcock and Singer [9], who studied molten potassium chloride. Computer experiments have since then been carried out for a large number of ionic systems, both in the liquid and solid state, and several excellent reviews have been written on the method and its applications. We shall therefore give only a brief resumé on the molecular dynamics method and refer the reader to the article by Sangster and Dixon [10] for details.

The molecular dynamics technique corresponds to the study of the micro-canonical ensemble for a model system. The system consists of N particles, enclosed in a cubic box of volume $V=L^3$, interacting with each other through a potential function $\Phi(\underset{\sim}{r}_1,\ldots\underset{\sim}{r}_N)$, which is generally assumed to be pair wise additive. Thus:

$$\Phi(\underset{\sim}{r}_1,\ldots\underset{\sim}{r}_N) = \sum_{i=1}^{N-1} \sum_{j=i+1}^{N} \phi(\underset{\sim}{r}_{ij}) \qquad (2.1)$$

In order to minimize the effects of the smallness of the system (N is usually of 10^2-10^3) periodic boundary conditions are normally used; i.e. the central box is surrounded on all sides by replicas of itself.

The calculation starts by allocating positions $\underset{\sim}{r}_i$ and velocities $\underset{\sim}{v}_i$ to the particles at time t=0. The force on each particle can then be evaluated from:

$$\underset{\sim}{F}_i = -\nabla \sum_{j\neq i} \phi(\underset{\sim}{r}_{ij}) \tag{2.2}$$

where the sum is over all particles in the central box and their images in neighbouring cells. When $\phi(\underset{\sim}{r})$ decays sufficiently fast, as is the case for most pair potentials, one can, without any significant loss in accuracy, neglect all interactions between pairs separated by a distance greater than some cut-off radius r_c. For convenience one usually chooses $r_c \leq \frac{1}{2}L$. Having calculated the forces the equations of motion are now integrated numerically by some suitable algorithm to yield the positions $\underset{\sim}{r}_i'$ and velocities $\underset{\sim}{v}_i'$ a short time Δt later. By repeating this procedure the dynamic state of the system is determined as a function of time. Equilibrium properties of the system are calculated by averaging over a sufficiently long time interval. Furthermore, since one generates a dynamic sequence of states, time-dependent phenomena can be studied. An extensive list of physical quantities which can be obtained from a molecular dynamics experiment is given in ref. 10.

In an ionic system the pair potential $\phi(\underset{\sim}{r})$ contains a Coulomb term proportional to r^{-1}. The slow decay of this term makes it impossible to truncate $\phi(\underset{\sim}{r})$ and we have to evaluate the full lattice sum. Techniques for obtaining rapidly convergent expressions for the Coulomb energy of a periodic system are discussed elsewhere in this volume [2]. For the case in which the periodic array of charges makes a macroscopically large sphere surrounded by a medium of dielectric constant ε' the expression for the electrostatic energy U_c can be written as [1,2]

$$U_c = \frac{1}{2} \sum_n \sum_{i=1}^{N} \sum_{j=1}^{N}{}' q_i q_j e^2 \frac{\operatorname{erfc}(\alpha|\underset{\sim}{r}_{ij\underset{\sim}{n}}|)}{|\underset{\sim}{r}_{ij\underset{\sim}{n}}|} + \frac{e^2}{\pi V} \sum_{\underset{\sim}{k}}{}' \frac{\exp(-\pi^2|\underset{\sim}{k}|^2/\alpha^2)}{|\underset{\sim}{k}|^2} \rho_{\underset{\sim}{k}} \rho_{-\underset{\sim}{k}}$$

$$+ \frac{\alpha e^2}{\sqrt{\pi}} \sum_i q_i^2 + \frac{2\pi e^2}{(2\varepsilon'+1)} \frac{\underset{\sim}{M}_p^2}{V} \tag{2.3}$$

Here q_i denotes the charge of particle i in units of the elementary charge e, $\underset{\sim}{r}_{ij\underset{\sim}{n}} = \underset{\sim}{r}_i - \underset{\sim}{r}_j + \underset{\sim}{n}L$. The sum over $\underset{\sim}{n}$ is over all translation vectors of the simple cubic lattice. The prime denotes that for $\underset{\sim}{n}=(0,0,0)$ the term i=j is omitted. erfc is the complementary error function. The sum over $\underset{\sim}{k}$ runs over all reciprocal lattice vectors except $\underset{\sim}{k}=(0,0,0)$ as indicated by the prime. $\rho_{\underset{\sim}{k}}$ is the Fourier-transform of the charge density:

$$\rho_{\underset{\sim}{k}} = \sum_j q_j \exp(2\pi i \underset{\sim}{k}.\underset{\sim}{r}) \tag{2.4}$$

V denotes the volume of the periodic cell and α is a disposable parameter governing the rate of convergence of the two series in eq. 2.3.

The last term in eq. 2.3 is the most interesting in that it depends explicitly on the dielectric constant ε' of the surrounding medium. The quantity $\underset{\sim}{M}_p$ is defined by:

$$\underset{\sim}{M}_p = \sum_i q_i \underset{\sim}{r}_i \tag{2.5}$$

It is of interest to see what happens if we translate a particle, say particle j, over a distance L in the x-direction. It is easily seen that the last term in eq. 2.3 then changes by an amount

$$\frac{2\pi e^2}{(2\varepsilon'+1)V} \{q_j^2 L^2 + 2q_j L M_p^x\}$$

where M_p^x denotes the x-component of $\underset{\sim}{M}_p$. Clearly this term, unlike the others in eq. 2.3, is not periodic. Hence, when a particle leaves the central cell in the course of a molecular dynamics experiment we may not replace it by its image in the cell as this would alter the Hamiltonian of the system. A moment of reflection will convince the reader that $\underset{\sim}{M}_p$ is related to the surface charge of the system. We must then make a distinction between $\underset{\sim}{M}_p$, which is the total moment of the given set of particles which are followed during the simulation experiment and whose positions may be in or out of the central cell, and $\underset{\sim}{\Lambda}$, which is the total moment of the central cell itself:

$$\underset{\sim}{\Lambda} = \sum_i q_i \underset{\sim}{r}_i' \tag{2.4}$$

where $\underset{\sim}{r}_i'$ is the position of the image of a particle within the central cell. $\underset{\sim}{\Lambda}$ is related to the dipole moment per unit volume, a bulk property.

Taking ε' to infinity will make the last term in eq. 2.3 disappear and we recover the well-known Ewald result.

3. INFLUENCE OF BOUNDARY CONDITIONS

The explicit dependence of the electrostatic energy of a periodic system on the dielectric constant ε' of the surrounding medium necessitates a detailed investigation into which properties are affected by a change in ε' and to what extent. A similar investigation for polar fluids [1,5] showed that in this case most properties are, to order 1/N, independent of ε'. The fluctuation in the total dipole moment of the simulation sample $\langle \underset{\sim}{M}^2 \rangle_{\varepsilon'}$, where $\langle\ \rangle_{\varepsilon'}$ denotes an ensemble average for the system whose surrounding medium has dielectric constant ε', was found to depend on ε' even in the thermodynamic limit. A change in ε' would alter $\langle \underset{\sim}{M}^2 \rangle_{\varepsilon'}$ in such a way that the same value of the dielectric constant ε is recovered provided ε is calculated from the general formula:

$$\frac{(\varepsilon-1)(2\varepsilon'+1)}{3(2\varepsilon'+\varepsilon)} = \frac{4\pi}{9k_BT}\,\frac{\langle \underset{\sim}{M}^2 \rangle_{\varepsilon'}}{V} \tag{3.1}$$

and proper account is taken of the boundary conditions in the simulation experiment. Similarly we may expect the quantity $\langle \underset{\sim}{M}_p^2 \rangle_{\varepsilon'}$ to depend strongly on ε' in the simulation of ionic systems. An expression giving the change in $\langle \underset{\sim}{M}_p^2 \rangle_{\varepsilon'}$ due to a change in ε' has recently been derived by Smith et al.[12] using perturbation theory. When we are dealing with an ionic liquid we can derive an expression for $\langle \underset{\sim}{M}_p^2 \rangle_{\varepsilon'}$ as a function of ε' by an electrostatic argument. The derivation is a good example of the connection between quantities obtained from a simulation experiment and their macroscopic counterparts.

We seek to derive an expression for the polarization induced in the system sketched in fig. 1 when a small constant electric field is applied. The spherical region of radius a (region I) contains the ionic fluid. It is surrounded by a much larger spherical shell with outer radius b of uniform dielectric with dielectric constant ε' (region II). The outside region (III) is vacuum. In our derivation we shall take a,b to infinity in such a way that a/b→0, as this is the system for which the Hamiltonian 2.3 is valid. We denote the constant external field by $\underset{\sim}{E}_e$ and take the polar axis in the direction of $\underset{\sim}{E}_e$. The only acceptable solutions of Laplace's equation for the potential Φ in regions II and III are:

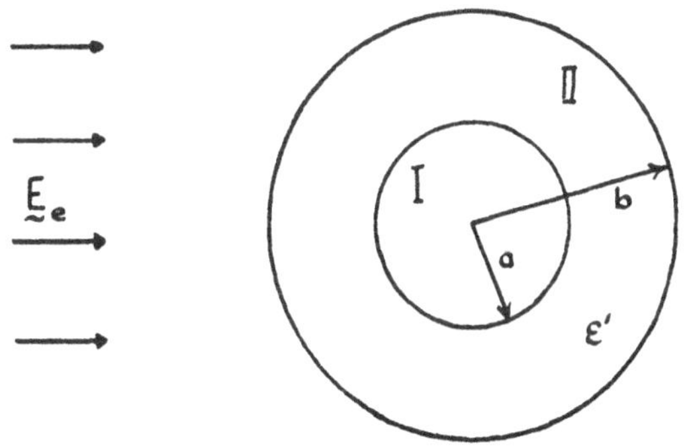

Fig. 1. Geometry used for the calculation of the polarization in an ionic fluid.

$$\Phi_{III}(r,\theta) = \frac{AE_e}{r^2} \cos\theta - E_e r \cos\theta \tag{3.2a}$$

$$\Phi_{II}(r,\theta) = \frac{BE_e}{r^2} \cos\theta + CE_e r \cos\theta \tag{3.2b}$$

where A, B and C are constants to be determined from the boundary conditions and $E_e = |\tilde{E}_e|$. For sufficiently small fields the potential in region I obeys the linearized Poisson-Boltzman equation:

$$\nabla^2 \Phi_I(r,\theta) = k_D^2 \Phi_I(r,\theta) \tag{3.3}$$

where k_D^2 is the inverse Debye screening length:

$$k_D^2 = \frac{4\pi e^2}{k_B T} \sum_\alpha q_\alpha^2 \rho_\alpha \tag{3.4}$$

with ρ_α denoting the density of species α and q_α its charge. It is easily verified that the only solution of 3.3, which does not go to infinity at the origin and has the appropriate symmetry, may be written as:

$$\Phi_I(r,\theta) = DE_e j_1(ik_D r) \cos\theta \tag{3.5}$$

$j_1(x)$ is the spherical Bessel function of order 1:

$$j_1(x) = \frac{\sin x}{x^2} - \frac{\cos x}{x} \tag{3.6}$$

and D is a constant. The boundary conditions require the continuity of $\Phi(r,\theta)$ and $\varepsilon\partial\Phi(r,\theta)/\partial r$ at r=a and b yielding four equations for the constants A, B,C and D. Solving for D we find, in the limit $a/b \to 0$:

$$D = \frac{q\varepsilon'}{(\varepsilon'+2)} \Delta \tag{3.7}$$

with

$$\Delta = -a(ik_D j_1'(ik_D a) + 2\varepsilon' j_1(ik_D a))^{-1} \underset{\text{large } a}{\sim} \frac{ia}{\sinh k_D a} \tag{3.8}$$

The charge distribution $Q(r,\theta)$ in region I is then obtained from Poisson's equation:

$$\nabla^2 \Phi_I(r,\theta) = 4\pi e Q(r,\theta) \tag{3.9}$$

with the aid of eq. 3.3 we find:

$$Q(r,\theta) = \frac{q\varepsilon'}{4\pi(\varepsilon'+2)} k_D^2 E_e j_1(ik_D r) \cos\theta \tag{3.10}$$

The polarization $\underset{\sim}{P}$ induced in region I is obtained from:

$$\underset{\sim}{P} = \frac{e}{V} \int \underset{\sim}{r}\, Q(\underset{\sim}{r})\, d\underset{\sim}{r} \tag{3.11}$$

with V is the volume of the system. Evidently $\underset{\sim}{P}$ and $\underset{\sim}{E}_e$ are parallel vectors, so that the magnitude of $\underset{\sim}{P}$ is given by:

$$P = \frac{3e}{4\pi a^3} \int r\cos\theta Q(r,\theta)\, dr \tag{3.12}$$

The evaluation of the integral is straight forward. In the limit $a \to \infty$ we find:

$$\underset{\sim}{P} = \frac{9}{4\pi} \frac{\varepsilon'}{(\varepsilon'+2)} \underset{\sim}{E}_e \tag{3.13}$$

Eq. 3.13 could have been derived more directly by treating the ionic fluid in region I as a perfect conductor. Such an approach is not always justified however.

We proceed by calculating $\underset{\sim}{P}$ using statistical mechanics. The field $\underset{\sim}{E}$ acting on the ionic fluid in region I can be calculated by assuming the region to be empty. It is then straight forward to show that:

$$\underset{\sim}{E} = \frac{9\varepsilon'}{(\varepsilon'+2)(2\varepsilon'+1)} \underset{\sim}{E}_e \tag{3.14}$$

The induced polarization in region I due to the field $\underset{\sim}{E}$ is:

$$P = \frac{1}{V} \frac{\int..\int d\underset{\sim}{r}_1..d\underset{\sim}{r}_N (e\sum_i q_i \underset{\sim}{r}_i) \exp(-\beta H_o + \beta \underset{\sim}{E}.e\sum_i q_i \underset{\sim}{r}_i)}{\int..\int d\underset{\sim}{r}_1...d\underset{\sim}{r}_N \exp(-\beta H_o + \beta \underset{\sim}{E}.e\sum_i q_i \underset{\sim}{r}_i)} \tag{3.15}$$

Expanding numerator and denominator in terms of $\underset{\sim}{E}$ retaining terms linear in the field only we have, for isotropic systems

$$\underset{\sim}{P} = \frac{1}{3} \beta e^2 E \frac{\langle \underset{\sim}{M}^2 \rangle}{V} \tag{3.16}$$

Combining eq. 3.13, 3.14 and 3.15 we finally obtain:

$$\frac{\langle \underset{\sim}{M}^2 \rangle}{V} = \frac{3}{4\pi} \frac{(2\varepsilon'+1)}{e^2} k_B T \tag{3.17}$$

In a simulation experiment it is then normally assumed that the quantity $\langle \underset{\sim}{M}^2 \rangle/V$ can be approximated by evaluating it over the simulation sample. In this case we must indentify the left hand side of eq. 3.17 with $\langle \underset{\sim}{M}_p \rangle$, rather than the mean square moment of the cell itself, since both the polarization $\underset{\sim}{P}$ and the quantity $M_{\sim p}$ are related to the surface charge of the system. Thus:

$$\frac{\langle M_{\sim p}^2\rangle_{\varepsilon'}}{V} = \frac{3k_BT}{4\pi e^2}\,(2\varepsilon'+1) \tag{3.18}$$

Smith et al. [12] derived the following expression for the change in $\langle M_{\sim p}^2\rangle$ due to a change in ε':

$$\langle M_{\sim p}^2\rangle_{\varepsilon''} = \frac{\langle M_{\sim p}^2\rangle_{\varepsilon'}}{1-\lambda(\varepsilon'',\varepsilon')\langle M_{\sim p}^2\rangle_{\varepsilon'}} \tag{3.19}$$

where the parameter $\lambda(\varepsilon'',\varepsilon')$ is given by:

$$\lambda(\varepsilon'',\varepsilon') = \frac{4\pi e^2}{9k_BTV}\left\{\frac{2(\varepsilon''-1)}{2\varepsilon''+1} - \frac{2(\varepsilon'-1)}{2\varepsilon'+1}\right\} \tag{3.20}$$

Substituting 3.18 for $\langle M_{\sim p}^2\rangle_{\varepsilon'}$ and $\langle M_{\sim p}^2\rangle_{\varepsilon''}$ yields an identity.

For the average value of the energy associated with the term explicitly dependent on ε' in the Hamiltonian 2.3 we get:

$$\frac{2\pi e^2}{2\varepsilon'+1}\,\frac{\langle M_{\sim p}^2\rangle_{\varepsilon'}}{V} = \frac{3}{2}\,k_BT \tag{3.21}$$

Eq. 3.21 shows that the second term in 2.3 only adds a small amount, independent of the number of particles, to the energy of the system and has no effect on the thermodynamics. Moreover, this is precisely the energy of a single classical harmonic oscillator, in line with the quadratic dependence of the energy on $M_{\sim p}$.

Finally we note that the eq. 3.18 and 3.21 have been verified in a molecular dynamics experiment on a simple molten salt [13].

Let us now turn to another quantity, the long wave-length behaviour of the charge structure factor $S_{qq}(k)$. In this case we seek to derive an expression for the polarization due to a spatially varying field. Let us assume the field acting on the system of fig. 1 is given by:

$$\underset{\sim}{E}_e = i\underset{\sim}{k}\,\Phi_o\,e^{i\underset{\sim}{k}\cdot\underset{\sim}{r}} \tag{3.22}$$

corresponding to an external potential of the form:

$$\Phi_e(\underset{\sim}{r}) = \Phi_o\,e^{i\underset{\sim}{k}\cdot\underset{\sim}{r}} \tag{3.23}$$

The charge distribution giving rise to such a potential field may be written as:

$$Q_e(\underset{\sim}{r}) = \frac{k^2\Phi_o}{4\pi e} e^{i\underset{\sim}{k}.\underset{\sim}{r}} \tag{3.24}$$

The equations describing the behaviour of the potential are:

$$\nabla^2\Phi(\underset{\sim}{r}) = -\frac{4\pi e Q_e(r)}{\varepsilon} \tag{3.25}$$

for regions II ($\varepsilon=\varepsilon'$) and III ($\varepsilon=1$), and

$$\nabla^2\Phi(\underset{\sim}{r}) - k_D^2\,\Phi(\underset{\sim}{r}) + k^2\Phi_o\, e^{i\underset{\sim}{k}.\underset{\sim}{r}} = 0 \tag{3.26}$$

for region I. Eq. 3.26 is a direct extension of the linearized Poisson-Boltzman equation 3.3 for the case in which an external charge distribution is present. The solutions of eq. 3.25 and 3.26 may be written as:

$$\Phi_{III}(r,\theta) = \sum_{\ell=0}^{\infty} \frac{A_\ell}{r^{\ell+1}} P_\ell(\cos\theta) + \Phi_o e^{ikr\cos\theta} \tag{3.27}$$

$$\Phi_{II}(r,\theta) = \sum_{\ell=0}^{\infty} \left\{\frac{B_\ell}{r^{\ell+1}} + B'_\ell r^\ell\right\} P_\ell(\cos\theta) + \frac{\Phi_o e^{ikr\cos\theta}}{\varepsilon'} \tag{3.28}$$

$$\Phi_I(r,\theta) = \sum_{\ell=0}^{\infty} C_\ell j_\ell\,(ik_D r)\, P_\ell(\cos\theta) + \frac{k^2}{k^2+k_D^2}\,\Phi_o e^{ikr\cos\theta} \tag{3.29}$$

$P_\ell(\cos\theta)$ denotes the Legendre polynomial of order ℓ, $j_\ell(x)$ the spherical Bessel function of order ℓ (Abramowitz and Stegun [14]). The constants A_ℓ, B_ℓ, B'_ℓ and C_ℓ are determined by matching the values of $\Phi(r,\theta)$ and $\varepsilon\partial\Phi(r,\theta)/\partial r$ at the boundaries of the regions. This is most conveniently done by using Gegenbauer's expansion for $e^{ikr\cos\theta}$ [14]:

$$e^{ikr\cos\theta} = \sum_{\ell=0}^{\infty} (2\ell+1)\, i^\ell j_\ell\,(kr)\, P_\ell(\cos\theta) \tag{3.30}$$

After several pages of tedious but straight forward algebra one finds in the limit $a,b \to \infty$, $a/b \to 0$:

$$C_\ell = \frac{a\Delta_\ell}{\cos(ik_D a - \frac{1}{2}\ell\pi)} \tag{3.31}$$

In deriving 3.31 we have used the asymptotic expansion for $j_\ell(z)$ [14]:

$$j_\ell(z) \sim \frac{\sin(z-\frac{1}{2}\ell\pi)}{z} \tag{3.32}$$

and Δ_ℓ is given by:

$$\Delta_\ell = (2\ell+1)\, i^\ell k\, \Phi_o j'_\ell(ka) \tag{3.33}$$

the prime denoting differentiation of the function with respect to its argument.

The calculation of the polarization proceeds along the same lines as in the preceeding example. We find the charge distribution induced in region I and calculate $\underset{\sim}{P}$ according to eq. 3.19. Clearly $\underset{\sim}{P}$ is parallel to $\underset{\sim}{k}$, so that its magnitude is given by 3.12. It is then easily seen that, because of the orthogonality of the Legendre polynomials, only the term with ℓ=1 survives. We thus obtain:

$$\underset{\sim}{P} = \frac{3}{4\pi}\, i\underset{\sim}{k}\, \Phi_o j'_1(ka) \underset{k\to 0}{\sim} -\frac{i\Phi_o}{4\pi}\, \underset{\sim}{k} \tag{3.34}$$

The potential acting on the charges in region I is obtained by solving the boundary value problem with the applied field of eq. 3.22 and treating region I as a spherical cavity:

$$\Phi_I^C(r,\theta) = \sum_{\ell=0}^{\infty} \frac{(\varepsilon'-1)(\ell+1)}{(\varepsilon'(\ell+1)+\ell)}\, \frac{\Gamma_\ell}{a^\ell}\, j_\ell(kr)\, P_\ell(\cos\theta) + \Phi_o e^{ikr\cos\theta} \tag{3.35}$$

with

$$\Gamma_\ell = (2\ell+1)\, i^\ell \Phi_o j_\ell(ka) \tag{3.36}$$

The polarization $\underset{\sim}{P}$ is then calculated as:

$$\underset{\sim}{P} = \frac{e}{V}\int_V d\underset{\sim}{r} \sum_\alpha q_\alpha \underset{\sim}{r}\, \rho_\alpha^{(1)}(\underset{\sim}{r};k) \tag{3.37}$$

where $\rho_\alpha^{(1)}(\underset{\sim}{r};k)$ is the single particle density of species α:

$$\rho_\alpha^{(1)}(\underset{\sim}{r};k) = \frac{\frac{1}{(N_\alpha-1)!N_\beta!}\int..\int d2_\alpha \ldots dN_\alpha d1_\beta \ldots dN_\beta \exp(-\beta H_o - \beta\sum_i q_i \Phi_I^C(r_i))}{\frac{1}{N_\alpha!N_\beta!}\int..\int d1_\alpha \ldots dN_\alpha d1_\beta \ldots dN_\beta \exp(-\beta H_o - \beta\sum_i q_i \Phi_I^C(\underset{\sim}{r}_i))} \tag{3.38}$$

For small k we write:

$$\rho_\alpha^{(1)}(\underset{\sim}{r};k) = \rho_\alpha^{(1)}(\underset{\sim}{r};0) + k\frac{\partial}{\partial k}\rho_\alpha^{(1)}(\underset{\sim}{r};k{=}0) + O(k^2) \tag{3.39}$$

and obtain after some algebra:

$$\underset{\sim}{P} = \frac{i\beta\Phi_o}{6}\underset{\sim}{k}\sum_\alpha\sum_\beta q_\alpha q_\beta\rho_\alpha\rho_\beta\int d\underset{\sim}{r}\, r^2\, h_{\alpha\beta}(r) \tag{3.40}$$

where $h_{\alpha\beta}(r)$ is the total correlation function for ionic species α and β.

Combining eq. 3.40 and 3.34 we arrive at the well known Stillinger-Lovett result [15]:

$$\frac{1}{\sum_\alpha q_\alpha^2\rho_\alpha}\sum_\alpha\sum_\beta q_\alpha q_\beta\rho_\alpha\rho_\beta\int d\underset{\sim}{r}\, r^2\, h_{\alpha\beta}(r) = -\frac{6}{k_D^2} \tag{3.41}$$

The left hard side of 3.41 is essentially the second moment of the charge structure factor $S_{qq}(k)$:

$$S_{qq}(k) = \frac{1}{N}\left\langle\sum_j\sum_\ell q_j q_\ell \exp(i\underset{\sim}{k}\cdot\underset{\sim}{r}_{j\ell})\right\rangle \tag{3.42}$$

so that the long wavelength behaviour of $S_{qq}(k)$ is given by:

$$S_{qq}(k) \underset{k\to 0}{\sim} \frac{k^2}{4\pi\beta e^2\rho} \tag{3.43}$$

ρ being the total density of the system $\rho = \sum_\alpha \rho_\alpha$.

We note that this result is independent of ε'. For those enjoying the algebra the derivation given above could be extended to yield higher moments of $S_{qq}(k)$. These moments will show some dependence on ε', to order $1/N$. However, in going beyond the linear regime the validity of the linearized Poisson-Boltzmann becomes questionable.

The periodic boundary conditions impose certain restrictions on the calculation of $S_{qq}(k)$ in a simulation experiment. We can only calculate the structure factor for wave vectors which coincide with reciprocal lattice vectors of the periodic system; i.e. we must have $\underset{\sim}{k} = \frac{2\pi}{L}(\nu_1,\nu_2,\nu_3)$ - the ν-'s being integers.

Expanding 3.42 we find

$$S_{qq}(k) \underset{k\to 0}{\sim} \frac{k^2}{3N}\left\langle\left(\sum_i q_i\underset{\sim}{r}_i\right)^2\right\rangle \tag{3.44}$$

Because of the periodicity we must now take the sum in 3.44 over the particles in the unit cell and thus identify the sum with $\tilde{\Lambda}$ rather than $\tilde{M}_p$ (see eq. 2.4). It then follows that:

$$\langle \tilde{\Lambda}^2 \rangle = k^2 V / 4\pi\beta e^2 \tag{3.45}$$

Thus the mean square moment of the unit cell is independent of ε'. This is not unexpected since $\langle \tilde{\Lambda}^2 \rangle$ is a bulk quantity. Eq. 3.45 has been verified in a simulation experiment [13].

So far we have only been concerned with the static properties of the system. Let us now turn to the dynamic properties and see what effect a change in ε' has on some of these. One effect, already implicit in the Hamiltonian 2.3, is immediately clear from 3.18. The time-dependence of $\tilde{M}_p$ is related to the zero frequency conductivity σ of the system via the Einstein relation:

$$\sigma = \frac{\beta e^2}{3V} \lim_{t\to\infty} \frac{\langle [\tilde{M}_p(t) - \tilde{M}_p(0)]^2 \rangle}{6t} \tag{3.46}$$

It follows at once that $\sigma=0$ for every finite value of ε'. Only for $\varepsilon'=\infty$ can be have $\sigma\neq 0$. This is not surprising since a finite ionic system acquires a surface charge when placed in an external field and no currents develop. Only when it is surrounded by a conductor ($\varepsilon'=\infty$) will an electric field give rise to a current. It follows that the behaviour of the charge-current correlation function $J(t)$, defined by:

$$J(t) = \frac{\langle \sum_i q_i \tilde{v}_i(t) \cdot \sum_i q_i \tilde{v}_i(0) \rangle}{\langle (\sum_i q_i \tilde{v}_i)^2 \rangle} \tag{3.47}$$

will be affected by the nature of the surrounding medium. This is clearly shown in fig. 2, where we have plotted $J(t)$ for a simple molten salt with $\varepsilon'=1$ and $\varepsilon'=\infty$ respectively [13,14]. Evidently the second term in the Hamiltonian 2.3 has a dramatic effect on the behaviour of $J(t)$. For $\varepsilon'=1$ $J(t)$ shows a strong oscillatory character with frequency $\omega \simeq 2\omega_p$ where ω_p is the plasma frequency, whereas for $\varepsilon'=\infty$ it is strongly damped. To analyse this phenomena one has to extend the methods developed for the static case to the time domain. Such an analysis will be given elsewhere.

Simulation experiments have further shown that other dynamic quantities, such as diffusion coefficients and local currents, are independent of ε' [13]. We may then conclude that simulations using eq. 2.3 for the calculation of the energy of an ionic system give bulk properties which are, to order $1/N$, independent of ε'. The use of eq. 2.3 makes it possible to calculate other quantities,

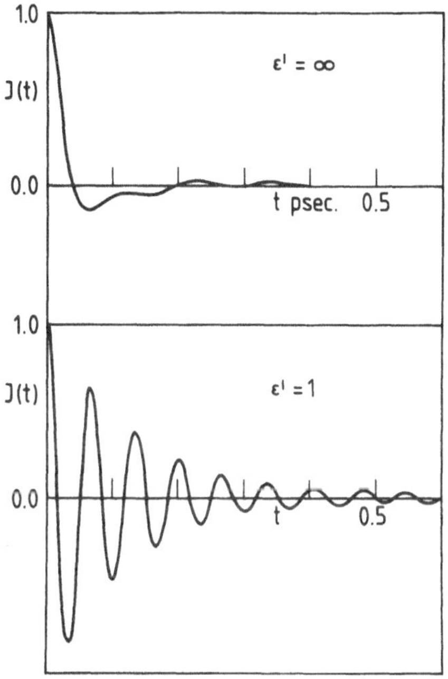

Fig. 2. Dependence of the current autocorrelation function on boundary conditions.

such as the surface charge acquired in an external field, from macroscopic arguments and compare these with the relevant quantities obtained from a simulation experiment.

REFERENCES

1. S.W. de Leeuw, J.W. Perram and E.R. Smith, Proc. Royal Soc. London A 373, 27 (1980).
2. E.R. Smith, this volume.
3. B.U. Felderhof, Physica 95A, 572 (1979).
4. B.U. Felderhof, Physica 101A, 275 (1980).
5. E.L. Pollock and B. Alder, Physica 102A, 1 (1980).
6. S.W. de Leeuw, J.W. Perram and E.R. Smith, Proc. Royal Soc. A 373, 57 (1980).
7. D. Levesque, G.N. Patey and J.J. Weis, Mol. Phys. 34, 1077 (1977).
8. S.G. Brush, H.L. Sahlin and E. Teller, J. Chem. Phys. 45, 2102 (1966).
9. L.V. Woodcock and K. Singer, Trans. Faraday Soc. 67, 12 (1971)
10. M.J.L. Sangster and M. Dixon, Adv. Phys. 25, 247 (1976).
11. P. Ewald, Ann. Phys. 64, 253 (1921).
12. E.R. Smith, C.S. Hoskins and C.C. Wright, Mol. Phys. (in press).
13. S.W. de Leeuw and J.W. Perram, Physica A (in press).
14. M. Abramowitz and I.A. Stegun, "Handbook of Mathematical Functions," New York, Dover, 1964.

INTRODUCTION TO MONTE CARLO SIMULATION TECHNIQUES

David Adams

Department of Chemistry
University of Southampton
Southampton SO9 5NH, England

INTRODUCTION: THE NATURE OF THE PROBLEM

Starting from the assumption that matter, for our purposes, consists of interacting particles obeying classical mechanics, and using the postulates of statistical mechanics, one can model any specific material as a system of particles provided one knows what the interactions between the particles are. However, whatever interactions are chosen the integrals that it is necessary to solve are formidable. For example, the average potential energy is

$$\langle U \rangle = \int U(\underline{r}^N) p(\underline{r}^N) d\underline{r}^N, \tag{1}$$

where the probability density for a configuration of N distinguishable particles, $\underline{r}^N \equiv (\underline{r}_1, \underline{r}_2, \ldots., \underline{r}_N)$ is

$$p(\underline{r}^N) = \exp\left[-U(\underline{r}^N)/kT\right]/Q_N, \tag{2}$$

where the configurational integral Q_N is

$$Q_N = \int \exp\left[-U(\underline{r}^N)/kT\right] d\underline{r}^N. \tag{3}$$

There are two reasons why these integrals are particularly difficult. The first reason is that ideally we would like answers, not for a small number of particles, but in the thermodynamic limit $N \to \infty$ and $V \to \infty$ while the density, $N/V = \rho$, remains constant. At the very least N will be, say, thirty and the integrals will be ninety dimensional. Usually, N will be in the hundreds!

Obviously such integrals as equation (3) cannot be performed analytically unless $U(\underline{r}^N)$ is trivially simple. Either one turns to approximations which are liable to be drastic for the liquid and solid phase, or one tries numerical integration. Now computers are excellent for solving one or two dimensional integrals by such methods as Simpson's rule or Gaussian quadrature. However, a ninety or more dimensional integral is just not practical with these methods. Suppose that we used quadrature and needed m evaluations of the integral to obtain a reasonably accurate value in order to integrate out one dimension. Then to perform the entire 3N dimensional integral we would need to calculate $\exp[-U(\underline{r}^N)/kT]$ at no less than m^{3N} points in 3N dimensional space. Even were m and N as small as 2 and 30 respectively this would still be a very long calculation indeed. If m and N were substantially larger, which we should very much prefer, then the Universe is not old enough for the calculation to have been performed! There is a standard numerical technique for solving multidimensional integrals when ordinary quadrature is impractical. Instead of trying to obtain values of the integrand at every point on a regular grid we find a very much smaller number of values scattered through the multidimensional volume of integration. This will not give a very accurate value of the integral, but provided that the sample points have been chosen without bias the answer obtained will also be unbiased. By taking more and more samples of the integrand the accuracy can be made as high as required. The simple way of sampling to do this is to sample at randomly chosen points. This is Monte Carlo integration, the use of random numbers, of chance, to obtain a definite, correct (but necessarily approximate) value of an integral.[1]

However, there is a second reason why integrals such as equation (3) defy ordinary means of solution. The integrands are very badly behaved. This is because $U(\underline{r}^N)$ is a steep function of every pair of particle separations. As any two particles are moved closer together then $U(\underline{r}^N)$ will change rapidly and become large and positive. Moreover, we have the exponential of this rapidly varying function to deal with. Thus the integrand will vary wildly in value from one random sample to another, even when the samples are comparatively close together. Worse than that, it would be found that the vast majority of samples would have very small values of $\exp[-U(\underline{r}^N)/kT]$ but that just occasionally no particle would be too close to another and the Boltzmann weighting factor would be larger by many orders of magnitude. For the comparatively "easy" case of only 32 hard spheres at half the close packed density this will occur for only one sample in 10^{39}![2] Within the large multi-dimensional volume over which we are trying to integrate there are tiny regions where the Boltzmann weighting factor will be so large that the rest of the volume can be ignored. These regions, of course, correspond to the possible phases that the bulk material can occupy, liquid, crystalline solid and possibly others such as the amorphous solid and liquid crystal phases. The only way that Boltzmann weighted integrals over configuration space can be performed is to concentrate exclusively on these regions.

It is fairly easy to locate these regions, particularly the solid phase, but quite a different matter to take samples in an unbiased way. If true ensemble averages and structural information are to be obtained it is essential that the whole of the important regions of configuration space should be sampled. A guess, based on our preconceived notions of what constitutes a typical configuration, would be quite useless. We require a method which automatically picks out, randomly and with a known bias, configurations of high probability. The method introduced by Metropolis, Rosenbluth, Rosenbluth, Teller and Teller in 1953 does just this.[3]

METROPOLIS MONTE CARLO: THE METHOD

Their method has become so nearly universal a method of Monte Carlo sampling in statistical mechanics that it has become known simply as Monte Carlo. This is now leading to confusion as completely different random-number methods are also described as Monte Carlo. I shall use the expression Metropolis Monte Carlo. The elementary Monte Carlo method is to sample at random in an unbiased fashion and weight each configuration with $\exp[-U(\underline{r}^N)/kT]$. The Metropolis Monte Carlo method is to produce configurations at random but with probability proportional to $\exp[-U(\underline{r}^N)/kT]$ and count each one equally. This is done in the form of a random walk in configuration space so biased that the walk will pass through configuration $\underline{r}^N$ with probability $\exp[-U(\underline{r}^N)/kT]$. This sounds difficult to do, and indeed the theory of the method can appear rather forbidding[4-6], but in practice it is simple and is easy to programme on a computer. The type of random walk generated is often referred to as a Markov chain. This simply means that the chance of going from one configuration at the kth step of the walk to some other at the (k+1)th step depends on the kth configuration but not on the (k-1)th or earlier configurations.

The basic problem is how to choose the (k+1)th configuration having arrived at the kth. Let the probability of stepping from a configuration labelled i to another labelled j be p_{ij}. Then, obviously,

$$\left.\begin{array}{l} p_{ij} \geqslant 0 \\ \sum_j p_{ij} = 1 \end{array}\right\} \qquad (4)$$

where the sum over j includes j = i. The requirement of Boltzmann weighting gives

$$\sum_i \exp[-U(i)/kT]p_{ij} = \exp[-U(j)/kT]. \quad (5)$$

One - but not the only - way of satisfying equation (5) is to require

$$\exp[-U(i)/kT]p_{ij} = \exp[-U(j)/kT]p_{ji},$$

the condition of microscopic reversibility, which can be rewritten as

$$p_{ij} = p_{ji} \exp\{-[U(j)-U(i)]kT\}. \quad (6)$$

This does not tell us how to choose our steps but it is suggestive of the Metropolis method. This is in two stages. First a new, trial configuration (j) is generated by making a small change to the current configuration (i) and the potential energy of j is calculated. Then the trial configuration becomes the next step in the walk with probability $M[U(j)-U(i)]$ where M is the Metropolis function,

$$M(\delta U) = \min[1, \exp(-\delta U/kT)]. \quad (7)$$

If j is not to be the next step in the walk then the next step is instead the previous configuration once again. This last is necessary in order to satisfy equation (4). The procedure satisfies equation (6) because

$$M(\delta U)/M(-\delta U) = \exp(-\delta U/kT), \quad (8)$$

but one other condition must be fulfilled. The probability that the first stage chooses j starting from i must be identically equal to the probability that the first stage would choose i starting from j. The method of deciding whether or not to accept the trial configuration is illustrated in figure 1.

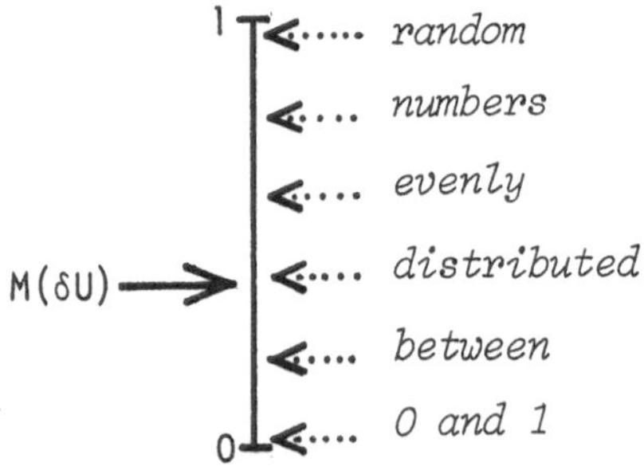

Figure 1 Graphic illustration of how a configuration is chosen with probability $M(\delta U)$. If $M(\delta U)$ is greater than the next random number then the trial configuration is accepted.

The trial configuration is adopted if the probability, $M(\delta U)$, is larger than a random number in the range (0,1]. If $\delta U = U(j)-U(i)$ is negative then $M(\delta U)$ is 1.0 and the trial configuration, j, is always adopted.

The common method of making the first stage is to choose one particle at random and give it a small random displacement so that its new position is equally likely to be at any point in a small cube centred on its old position. There is nothing special about a cube, it is just convenient to make a similar random change to each of the Cartesian coordinates. Metropolis et al[3] chose the particles in succession, rather than at random, and some simulators still do this. The relative merits of random choice or taking the particles one by one in order are not well understood, but there is probably little difference between them. The reason for moving just <u>one</u> particle at each step is cheapness. It is not necessary, you will note, to calculate U(j) at each step but only the change U(j)-U(i), and if only one particle has been moved then only the interactions of that one particle with the others has to be summed. Common practice is to adjust the maximum possible displacement so that ∿40-60% of trial configurations are accepted. The argument is that if the acceptance ratio is high, because the random displacements are very small, then the random walk will be too confined to a small region and will not properly sample all the available configuration space. If the acceptance ratio is low, because the random displacements are large, then the same configurations are being sampled repeatedly and the sampling efficiency is again poor. In order to adequately sample all the important regions of configuration space a very large number of steps is always necessary. Virtually all Metropolis Monte Carlo calculations have been in the range 10^5 to 10^7 steps and runs are getting longer with faster computers and more difficult problems.

One beautiful advantage of the Metropolis method is that whatever unlikely configuration is used to start the calculation the random walk will always tend towards a region of high Boltzmann weighting. The procedure is therefore to divide the calculation into two parts. The first "equilibration" part is necessary to ensure that the region to be sampled has a high Boltzmann weighting, and is not merely a consequence of the chosen starting configuration, which will either be a perfect lattice, which is a very convenient starting point, or a configuration from a previous calculation. The major part of the calculation follows when it is judged that the calculation has "equilibrated". It is only during this second, "production" part of the calculation that running sums of all the quantities for which ensemble averages are required are kept. Equation (1), for example, becomes:

$$\langle U \rangle = \frac{1}{n} \sum_{k=1}^{n} U(k) \qquad (9)$$

where there are n steps in the "production" part of the calculation and U(k) is the potential energy of the configuration at the kth step.

METROPOLIS MONTE CARLO: THE RESULTS

First let us consider what sort of system is suitable for Metropolis Monte Carlo simulation. Any one phase, vapour, liquid or solid presents no problems. It is normal to use periodic boundary conditions so that effectively a portion of the bulk material is simulated, rather than a small, isolated cluster, though those too can be studied. I shall leave the explanation of periodic boundary conditions to Professor Rahman (Chapter). It is possible also to study phase boundaries, and the most common method is to start with a thin ribbon of one phase with the other phase on both sides of it. This fits in with periodic boundary conditions so that a thin layer of infinite extent is simulated. One difficulty is that the layer must be very thin, a few Angstroms, and it must be inevitable that one surface affects the properties of the other. Most work so far has been on simple liquid/vapour interfaces. Julian Clarke's group at Manchester have published the only work on the surfaces of dense ionic systems known to me and they used the molecular dynamics method.[7] It is a general problem that only a very small volume can be simulated and two-phase regions, the critical point and any part of the phase diagram where the compressibility is high present special problems. There is no difficulty in doing the calculation but the results may bear little resemblance to the true properties in the thermodynamic limit! Outside such difficult areas the number of particles in the simulation should have little effect on the outcome though it is always wise to check this by repeating one or two calculations with a substantially different number of particles. The choice of the number of particles is often difficult, but for most purposes ∿100 should be regarded as the lower limit for condensed phases. The larger the number of particles the more expensive the calculation and a practical upper limit of ∿1000 applies in most cases. The largest calculations in three dimensions have so far not exceeded ten thousand particles, to the best of my knowledge. My own experience is that the number of steps necessary in the "equilibration" part increases more steeply than the number of particles, whereas the number of steps in the "production" part necessary for an acceptable accuracy seems independent of the size of the system.

The computer time required depends on the number of particles and on the complexity of the model potential, $U(\underline{r}^N)$. In most cases virtually all the computing time is taken by the calculation of δU at each step, and so the form of $U(\underline{r}^N)$ is very important. Omitting the possibility of angular dependence this may be expanded:

$$U(\underline{r}^N) = \sum_i \phi_i^{(1)}(\underline{r}_i) + \sum_{i<j}\sum \phi_{ij}^{(2)}(\underline{r}_i,\underline{r}_j) + \sum_{i<j<k}\sum\sum \phi_{ijk}^{(3)}(\underline{r}_i,\underline{r}_j,\underline{r}_k) \quad (10)$$

$$+ \ldots\ldots$$

The first set of terms, involving $\phi_i^{(1)}$, is some form of externally applied potential, and will most often be zero. An external electric field can be applied, but only when the system can reach an equilibrium. A field which would produce a steady-state current can only properly be treated using molecular dynamics. Normally the $\phi_{ij}^{(2)}$ are functions only of $|\underline{r}_i - \underline{r}_j|$ and are then ordinary potentials. In a simple salt there will be three distinct pair potentials, ϕ_{++}, ϕ_{+-} and ϕ_{--}. Calculating δU when only $\phi_{ij}^{(2)}$ terms are present involves a simple sum:

$$\delta U = \sum_{m\neq n} \{\phi_{mn}(|\underline{r}_m - \underline{r}_n'|) - \phi_{mn}(|\underline{r}_m - \underline{r}_n|)\} \quad (11)$$

where particle n has been moved from $\underline{r}_n$ to $\underline{r}_n'$ and m is summed over, at most, all other particles. In practice the evaluation of equation (11) is slightly more involved because each $|\underline{r}_m - \underline{r}_n|$ may have to be modified because of the periodic boundary conditions. When the triplet terms, $\phi_{ijk}^{(3)}$, are present then each calculation of δU will involve a double sum which takes $\sim N/2$ times as long to perform. This makes the calculation enormously expensive and virtually no calculations have been made involving triplet or higher terms. They have been used in one molecular dynamics simulation of a two dimensional fluid[8]. The effects of the polarizability of the ions or molecules on the potential energy can be written in the form of equation (10) but the higher n-tuple terms are important. The polarization of each molecule or ion is determined, at least in part, by the polarization of all the other molecules and ions. Fortunately, for polarizable ions and polar molecules the polarization can be determined with sufficient accuracy by an iterative method.[9] This is still an expensive calculation, but much less so than incorporating triplet terms explicitly, and with the arrival of array processing computers such work is now quite feasible. Molecular dynamics seems the natural method for this sort of work and I know of only one Metropolis Monte Carlo study.[10]

There is an important complication which makes the evaluation of the potential energy rather more complicated than equation (11) suggests. The model pair potentials we wish to use are generally defined out to infinity, whereas we have a small number of particles in the periodic cell and are severely limited in the number of interactions over which we can sum. When the pair potential falls rapidly with increasing separation this is no great problem. The calculation of each δU need only include interactions between near-neighbour particles and the pair potential can be truncated to zero at some cutoff radius, r_c. A correction for the omitted long-range interactions can be made approximately:

$$U_{ij}^{LRC} = \frac{1}{2}\,\frac{N_i N_j}{V}\,4\pi \int_{r_c}^{\infty} r^2 \phi_{ij}(r)\,dr \tag{12}$$

where volume V contains N_i of species i and N_j of species j. There is a corresponding long-range correction to the virial which can make a substantial contribution to the total computed pressure.

A very real problem arises when the pair potential does not fall rapidly to zero, when for example, it is the Coulombic potential between a pair of ions. However large the periodic cell the potential will always be much longer ranged than half the cell length, the customary r_c. The answer is to consider the infinite lattice of periodic cells and to sum over the entire lattice by using the Ewald sum method. This expresses the total Coulombic energy per cell in the form of rapidly convergent series. I will leave it to others (see Chapter 3) to explain the method and will try merely to explain why such a method is necessary. The cutoff

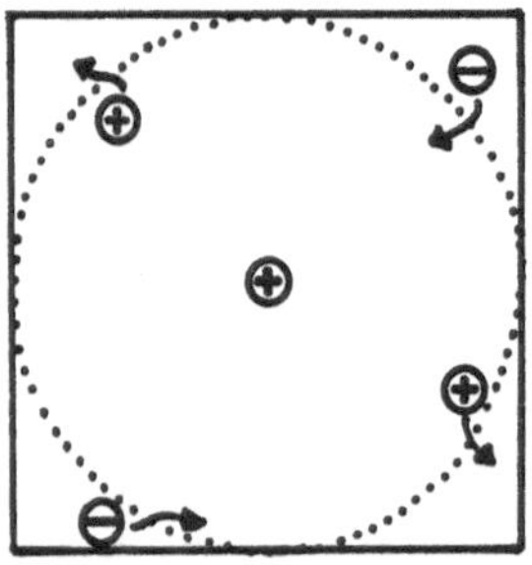

Figure 2 Cut-off sphere and nearest image cube of one ion in a simulation. With spherical truncation a considerable drop in energy is produced by the ions near the cut-off limit moving as shown. Summing over the entire nearest image cube still distorts the liquid structure as ions of opposite sign tend to crowd into the corners.

sphere around a positive ion and also its cube of nearest-neighbour ions are sketched in figure 2.

Lower energy configurations result by ions of the same sign crossing out of the truncation boundary while ions of the opposite sign cross in. This effect occurs around every ion and the resulting liquid structure is grossly distorted. It has been argued that the effect can be removed by summing over an electrically neutral region around each ion, that is, the cube of nearest neighbour particles.[5] However, this method is also far from satisfactory, the gross effect of a sudden cutoff is removed but even so configurations of lower energy will still result from ions of the same sign crowding into the corners of the truncation cube, as illustrated in figure 2. The cube shape of the boundary conditions tends to distort the fluid structure away from spherical symmetry to a cubic symmetry and curious and unreal configurations will sometimes result. There is no doubt that such configurations can occur but my explanation of them is not universally accepted. The spherical truncation method can be made to work with the addition of a suitable long-range correction - that of equation (12) is obviously not - but the results are disappointingly number dependent. My conclusion is that an Ewald method is best unless an unusually large number of ions is needed.[11]

The short-ranged part of the pair-potential never seems to cause problems. However complicated it may be it can always be tabulated in advance and rapidly determined by linear interpolation or simply by using the nearest tabulated value. A simple look-up method originated some years ago by McDonald and Singer[12] has much to recommend it. This holds particle coordinates as integers to permit rapid array look-ups and determination of nearest image distances. With modern computers, larger arrays and a finer grid of integer coordinates may be used.

Having discussed the types of system that can be studied we must now ask what information can and cannot be obtained from Metropolis Monte Carlo. The quantity that can be obtained the most accurately is the ensemble average potential energy, and the total energy is simply

$$\langle E\rangle = \frac{3}{2} NkT + \langle U\rangle \tag{13}$$

The separate ensemble average contributions of the various terms in the pair potentials, Coulombic, dispersion, short-range repulsion, etc., can also be computed with good reliability. Pressure is found from the ensemble average virial:

$$p = \frac{NkT}{V} - \frac{1}{3V} \left\langle \sum_{i<j} r \frac{\partial \phi_{ij}}{\partial r}\bigg|_{r_{ij}} \right\rangle \tag{14}$$

At low pressures this is the small difference between large terms and it is not possible to calculate pressure with any precision. Even for a simple fluid such as liquid argon the accuracy will be only to plus or minus a few tens of atmospheres and for molten salts this will be plus or minus a few hundred atmospheres. The specific heat can be obtained directly from

$$C_V \equiv \left(\frac{\partial E}{\partial T}\right)_V = \frac{3}{2}\,Nk + \frac{\langle U^2\rangle - \langle U\rangle^2}{NkT^2} \tag{15}$$

but this also involves the small difference between large quantities and usually the results obtained for C_V are only accurate to ±10% or even ±20% and are not worth quoting. Thermodynamic derivatives are more usefully obtained by making a series of calculations at a variety of densities and temperatures and then numerically differentiating.

In appropriate cases other static bulk properties such as dielectric constant or elastic moduli can be calculated also. Unfortunately, entropy, chemical potential and free energy are unobtainable. The reason is that a knowledge of the partition function, equation (3), is required for the configurational free energy,

$$A_c = -kT \ln Q_N, \tag{16}$$

to which the chemical potential and entropy are simply related. The random walk method of Metropolis et al does not give us the multi-dimensional volume of the important region of configuration space that it samples, and without that the integral of equation (3) cannot be evaluated even approximately. There are other means of obtaining the free energy and Dr. Quirke's lecture is concerned with these (Chapter 10).

During the course of the calculation all sorts of microscopic properties, such as radial distribution functions, mean-square displacements from lattice sites, etc., can also be calculated. I shall say little about these as the properties of interest will depend very much on both the system under investigation and the interests of the investigator. However, within the basic limitation of a small periodically repeating cell, <u>any</u> property that is a function only of the particle positions can be obtained. It is up to the investigator to devise interesting functions and include their calculation in his or her programme. The common way of computing a radial distribution function is to compute every (modified) pair separation once every few hundred or thousand steps and use these to construct a histogram of pair separations from which g(r) can later be obtained. It is not common practice to store the configurations produced during the calculation on magnetic tape for later analysis, as it is for molecular dynamics, but there is no good reason why this should not be done if there is any doubt as to whether the initial version of the programme computes everything that could be of interest. My own practice has always been to average everything that could possibly be useful, but all too often I have discovered something I have overlooked half-way through the calculation!

An estimate of the accuracy of the results obtained can be made by dividing the calculation into a small number of shorter walks each with its own subaverages. Standard statistical formulae can then be used to obtain an estimate of the standard deviation and standard error. However, this assumes that the subaverages are independent of each other and this will rarely be exactly true. Wood has discussed the problem of error estimation.[4] Additionally, it should be borne in mind that the reliability of the results obtained increases only as the square root of the length of the calculation and then only when the calculation is sufficiently long. Unfortunately, it is not always obvious when a calculation is of sufficient length. A short random walk may show an apparently good convergence, but only because it has not had the opportunity to properly sample more than a portion of the important regions of configuration space.

FORCE-BIASED AND SIMILAR METHODS

This can be a very awkward problem with some molecular liquids or even mixtures of simple spherical particles.[4] The random walk seems to "log-jam" in certain configurations and however long the calculation a proper sampling of configuration space will never occur. There is a general belief, which I share, that molecular dynamics is superior to Metropolis Monte Carlo in this respect, but this is not proven. What is true is that the simple Metropolis Monte Carlo method is capable of improvement and that this may make it superior to molecular dynamics as a means of sampling equilibrium configurations.

A simple example is the Metropolis Monte Carlo study of a 50/50 melt of NaCl and KCl. It was found that the statistics could be improved by including a special Monte Carlo step which attempted to swap over randomly chosen sodium and potassium ions.[13] Another simple case occurs when a quantity of interest depends more strongly on some particle positions than others. Then improved statistics can be obtained by more frequently choosing the important particles for random moves.[14,15]

Far more important than these rather rare cases is the recent, continuing development of more sophisticated random walks.

These satisfy equations (4) to (6) while making use of the forces acting on the particles in deciding the trial configurations.[16-19] The general idea is to make the calculations more like molecular dynamics.[16] It is too early to judge these methods properly, and anyway their efficacy depends on the system under study, but they do seem to be more efficient than Metropolis Monte Carlo, even for simple liquids[19], and are only slightly more complicated to programme. The Force-Biased method is claimed to give better sampling of liquid water configurations than molecular dynamics and even better methods may soon be devised. Rao and Berne[19]. comment on two methods so:

> *"In our view these two present algorithms represent only the beginning in the search for more rapidly convergent schemes, and smarter algorithms should soon be in the offing."*

My own, very limited experience of Force-Biasing applied to polar liquids has not been encouraging and suggests that molecular dynamics is more reliable, though I do not dispute that the Forced-Biased method is some improvement on Metropolis Monte Carlo. However, that improvement is not very great and long random walks will still be needed. The Force-Bias method was originally devised because Metropolis Monte Carlo and molecular dynamics gave different results for the potential energy of simulated liquid water.[16] It is now known that this difference was largely due to the use of different methods of summing the pair potentials and not to a failure of the Metropolis Monte Carlo calculation to adequately sample configuration space.

Force-Bias and related methods can be obtained by replacing equation (7) with the more general expression

$$\min\left[1,\ T(i|j)\exp(-\delta U/kT)/T(j|i)\right] \tag{17}$$

for the probability of accepting the trial configuration. $T(j|i)$ is the probability of the choice of trial configuration j given that the current configuration is i. The Metropolis method requires $T(i|j)=T(j|i)$ for all i and j. The hope is that a better choice of $T(i|j)$ can be made which will improve the sampling of configuration space. The original Force-Bias method used

$$T(j|i)\ \alpha\ \exp\left[\underline{F}(\underline{r})\cdot(\underline{r}'-\underline{r})/kT\right] \tag{18}$$

where $\underline{F}(\underline{r})$ is the force on the particle that is being moved from $\underline{r}$ to $\underline{r}'$. The method is more likely to choose to move the particle in the direction of the force acting on it and the larger the force the larger the displacement, up to the predetermined maximum displacement. The probability of accepting the trial configuration is biased in the opposite way so that equation (6) is fulfilled and Boltzmann weighted configurations are produced.

There may be more efficient schemes for implementing Force-Biased Monte Carlo but the following is at worst only a few percent slower than Metropolis Monte Carlo, and is applicable to any "smart" method.

1. Choose a particle at random. Sum both the forces acting on it and its energy of interaction with the other particles.
2. Make a trial displacement, as for Metropolis Monte Carlo.
3. Calculate $T(j|i)$ for this trial displacement.
4. If $T(j|i)$ is smaller than a random number in the range $(0,1]$ go back to step 2.
5. Calculate the force and potential of the particle at its trial position.
6. Accept the trial configuration with probability given by expression (17).

WHY MONTE CARLO?

This seems a good point to stop and ask the question, why use any Monte Carlo sampling method when molecular dynamics may give better sampling and enables the calculation of the huge range of time-dependent properties which are impossible to Monte Carlo? In fact researchers are increasingly using molecular dynamics. A Monte Carlo program, even with a sophisticated biasing scheme is less difficult to write, but a programme has only to be written once. Molecular dynamics and Monte Carlo calculations require comparable computing times, but molecular dynamics generally needs much more core store. However, this is increasingly less of a problem as ever larger computers become available.

Because there are none of the complications of choosing the best time step or trying to get the desired temperature by velocity scaling Monte Carlo is easier to perform. If it is intended to make many calculations at a series of V,T points to obtain the equation of state so that the properties of a model potential can be compared with experimental results, then the extra complications of molecular dynamics are rather bothersome and Monte Carlo is preferable.

There are also certain potentials, such as the charged hard sphere, which are particularly difficult for molecular dynamics but present no special problems for Monte Carlo sampling. This important potential has only been studied by Monte Carlo, as far as I am aware, because it does present so great a problem for molecular dynamics.

Apart from such an awkward case all the Monte Carlo calculations I have mentioned so far could have been done by molecular dynamics. I turn now to Monte Carlo calculations which are outside the scope of conventional molecular dynamics.

OTHER ENSEMBLES AND PHASE CHANGES

Very often experimental data does not cover a wide range of densities but is of the form of density, etc. at atmospheric pressure as a function of temperature. Excess properties of mixing are usually measured at constant pressure, and in general experiments are usually conducted at constant pressure, rather than constant volume. The Monte Carlo calculations I have described sample from the constant NVT ensemble and molecular dynamics is performed with constant number, volume and energy. However, it is possible to sample the constant NpT ensemble by Monte Carlo and obtain results for a set pressure and temperature.[4] This has been done with success for mixtures of simple liquids[20] and for crystalline and molten alkali halides[21]. It has also been used satisfactorily for molten mixtures of alkali halides[13], but failed to give adequate statistics in a similar series of calculations.[22]

The method has not been widely exploited and I think it deserves more attention as an alternative to conventional methods. The NpT ensemble partition function is a weighted sum over all NVT ensembles with the same temperature and number of particles:

$$\Delta(N,p,T) = \lambda\int_0^\infty \exp(-pV/kT)Z(N,V,T)dV \tag{19}$$

where the canonical partition function, Z, is given by

$$Z = \left(\frac{2\pi mkT}{h^2}\right)^{3N/2} \frac{1}{N!} Q_N \tag{20}$$

The factor 1/N! allows for the indistinguishability of the particles. The factor λ has dimensions of inverse volume, it does not enter the expressions for the ensemble averages.[4] The integral of equation (19) is taken over all ensembles of the same <u>shape</u>. Equations (19) and (20) are related to the thermodynamic properties by the bridging equation for the Gibb's free energy,

$$G(N,p,T) = -\,kT\ln\Delta \tag{21}$$

Expressions connecting thermodynamic properties with ensemble averages can be derived from equation (21) via the standard thermodynamic formulae. Metropolis Monte Carlo gives the ensemble average potential energy and the ensemble average volume from which the average density is known. The ensemble average virial can be computed to check that it agrees with the parametric pressure.

Different fluctuation formulae apply for different ensembles, and equation (15) for the specific heat at constant volume is now replaced by

$$C_p \equiv \left(\frac{\partial H}{\partial T}\right)_p = \frac{3}{2} Nk + \frac{<(U+pV)^2> - <U+pV>^2}{kT^2} \tag{22}$$

where H is the enthalpy:

$$H = \frac{3}{2} NkT + <U + pV> \tag{23}$$

Fluctuation formulae for the isothermal compressibility and the thermal expansivity can also be obtained from equation (21); they have been used in the Metropolis Monte Carlo study of alkali halide crystals at room temperature.[21] At higher temperatures the statistics will be poorer and the fluctuation formulae of little practical use unless the random walks are of exceptional length. With molten salts the reliability in the specific volume will typically be better than 1%, while the potential energy can be obtained to a reliability of $\sim$0.2%, comparable to the systematic error in the evaluation of the energy of individual configurations. However, mixtures of molten salts provide much greater problems.[13,21]

The implementation of NpT Metropolis sampling is only slightly more involved than NVT. At each step, in addition to moving a particle, a small random change is made in the length of the Monte Carlo cell. Naturally the change has to be equally likely to be positive or negative. This change alters slightly the separation between every pair of particles so that the total potential energy is altered. It is impracticable to resum all the pair interactions at each step but fortunately this is not necessary. Each term in the pair potential that is an inverse power of separation scales with the change in cell size, and the alteration in its contribution to the total energy is easily found. Terms which do not scale in this way can be approximated by a Taylor's series expansion about the cell length with good accuracy.[21] The trial configuration is adopted with probability $M(\delta W)$, where δW is the pseudo-Boltzmann weighting factor:[4,20]

$$\delta W = \delta U + p\delta V - NkT \ln \left[V(j)/V(i)\right] \tag{24}$$

where $\delta V = V(j)-V(i)$ is the change in cell volume. The origin of the first two terms is obvious, the third arises because Q_N has the dimensions of volume to the power N. Consider the case of a perfect gas, where $\delta U = 0$, and you will soon see that this is so.

NpT ensemble Monte Carlo was first used for the fluid of hard spheres[4] as it avoids the problem of having to extrapolate g(r) to contact to find the pressure. It is the ideal method of investigating systems of charged hard spheres, but has never been used for that purpose.

It is also possible to use the grand canonical ensemble with Metropolis Monte Carlo, a technique devised independently on at least three occasions[23-25]. In this case it is necessary to have special steps which add and remove particles from the Monte Carlo cell. This method can be used for the dense liquid at temperatures well above the triple point, but it is most successful for dilute systems. It has been used to obtain the liquid/vapour coexistence curve of the Lennard-Jones fluid[26], to study adsorption at the vapour/solid interface[27-29], and both bulk electrolyte solutions[30] and the diffuse double layer.[31] The method has been reviewed elsewhere.[2,32] I shall not go into details here as I think it unlikely to be suitable for ionic solids and their melts because the probabilities of successfully adding and removing ions will be extremely small.

I must now discuss a problem which I only hinted at earlier in this lecture: the determination of the thermodynamically stable phase. This can be difficult, unfortunately. The starting configuration often has a considerable influence on the outcome of the simulation. If there are a number of separate important regions of configuration space, each corresponding to a different phase, the random walk is unlikely to move from one to another. Much the same happens with molecular dynamics. If the starting configuration is liquid phase then the spontaneous crystallization to the solid phase is extremely unlikely, no matter how much more stable is the solid phase. Indeed, rapid cooling in a molecular dynamics experiment can produce an amorphous, glassy solid even for very simple fluids[33]. If the starting configuration is a perfect lattice then a liquid phase will not necessarily result just because it is the more stable phase, but a metastable solid phase may persist, or occasionally some weird semicrystalline structure may be produced because of the boundary conditions.[4] Again, liquid phase simulation sometimes results in negative pressures, that is, the metastable liquid remains stable and does not break into two phases. If, however, the simulation had started as two phases, as a ribbon of liquid with vapour on both sides, then this will persist also.

The persistence of a metastable phase can be either a bonus or a curse depending on the objectives of the investigator. It will be most extreme for first order phase changes using the NVT ensemble. Less has been done with 2nd order phase transitions, and this is something on which I am not well informed. My guess is that a second order transition from ordered to less ordered phase will occur reliably but that the reverse transition may give problems. NpT ensemble Metropolis Monte Carlo can be used to locate first order transitions approximately, but there can be considerable hysteresis. It is perfectly easy to use the method to study the liquid or solid phase under zero applied pressure even though the stable phase is a vapour of high volume. It has recently been claimed that the presence of hysteresis using the NpT ensemble proves a transition to be first order.[34]

Clearly, it can be difficult to locate a phase change better than very roughly, and it can even be difficult to decide the order of the transition. In any case, only in the thermodynamic limit is a transition sudden and sharp. There are two ways of determining the phase transitions of simulated systems more accurately. The first is to try and obtain both phases in coexistence in the periodic cell. This has been done for several liquid/vapour interfaces and is equally suitable for solid/vapour and possibly solid/liquid. The results are likely to be rather approximate, but a large improvement over trying to observe phase changes.

The alternative and more accurate method is to obtain free energy information for both phases and locate the unique conditions at each temperature at which the two phases have the same pressure and Gibbs free energy. The method of estimating free energy differences is the subject of Dr. Quirke's lecture (Chapter 10).

FINAL REMARKS

It is very tempting to identify the changes of the configuration as a calculation proceeds with the time evolution of the system. This is wrong, of course, but it is possible to infer some qualitative information about time dependent processes. Thus if a lattice structure remains stable but some particles show considerable mobility then it is reasonable to infer that those particles would diffuse in a molecular dynamics simulation while the others would not.

Many ways of incorporating random processes into the simulation of the time evolution of systems are now being developed,[35] and it can be very confusing that some of these are also referred to as Monte Carlo. One example is the use of Monte Carlo integration for modelling low pressure unimolecular reactions, where each step corresponds to the mean time between collisions.[36] Murch and Thorn use the expression Monte Carlo in their computer model of β"-alumina.[37] Though their method has similarities with Metropolis Monte Carlo it is in fact a <u>model</u> for the kinetics of the sodium ion in β"-alumina. Such model studies can be interesting and enlightening but ultimately they are only of value if their assumptions can be justified by more fundamental mechanical or statistical mechanical studies.

Dr. Quirke and I have concentrated on recent developments in Monte Carlo methods which have as yet seen little or no application to ionic solids. It is our hope that at least some will soon find a use in this field of research, there is plenty to do!

REFERENCES

1. J. M. Hammersley and D. C. Handscomb, "Monte Carlo Methods", Methuen, London (1964).
2. J. P. Valleau and G. M. Torrie, A Guide to Monte Carlo for Statistical Mechanics: 2. Byways, in: "Modern Theoretical Chemistry, Vol. 5A, Equilibrium Statistical Mechanics of Fluids", B. J. Berne, ed., Plenum, New York (1977).
3. N. Metropolis, A. W. Rosenbluth, M. N. Rosenbluth, A. H. Teller, and E. Teller, J. Chem. Phys. 21:1087 (1953).
4. W. W. Wood, Monte Carlo Studies of Simple Liquid Models, in: "The Physics of Simple Liquids", H. N. V. Temperley, J. S. Rowlinson, and G. S. Rushbrooke, eds., North-Holland, Amsterdam (1974).
5. J. P. Valleau and S. G. Whittington, A Guide to Monte Carlo for Statistical Mechanics: 1. Highways, in: "Modern Theoretical Chemistry, Vol. 5A, Equilibrium Statistical Mechanics of Fluids", B. J. Berne, ed., Plenum, New York (1977).
6. C. Zannoni, Computer Simulations, in: "The Molecular Physics of Liquid Crystals", G. R. Luckhurst and G. W. Gray, eds., Academic Press, London (1979).
7. D. M. Heyes, M. Barber and J. H. R. Clarke, J. C. S. Faraday II 73:1485 (1977); 75:1240, 1469 and 1484 (1979).
8. W. Schommers, Phys.Rev. A16: 327 (1977).
9. M. J. L. Sangster and M. Dixon, Adv.Phys. 25:247 (1976).
10. G. N. Patey, G. M. Torrie, and J. P. Valleau, J.Chem.Phys. 71:96 (1979).
11. D. J. Adams, Chem. Phys. Lett., 62:329 (1979); in: "The Problem of Long-Range Forces in the Computer Simulation of Condensed Media", D. Ceperely, ed., NRCC, Lawrence Berkeley Laboratory (1980).
12. I. R. McDonald and K. Singer, Disc. Faraday Soc. 43:40 (1967).
13. D. J. Adams and I. R. McDonald, Molec. Phys. 34:287 (1977).
14. D. R. Squire and W. G. Hoover, J. Chem. Phys. 50:701 (1969).
15. D. J. Adams and J. C. Rasaiah, Faraday Disc. 64:22 (1978)
16. C. Pangali, M. Rao, and B. J. Berne, Chem. Phys. Lett. 55:413 (1978).
17. P. J. Rossky, J. D. Doll, and H. L. Friedman, J. Chem. Phys. 69:4628 (1978).
18. M. Rao, C. Pangali, and B. J. Berne, Molec. Phys. 37:1773 (1979).

19. M. Rao and B. J. Berne, J. Chem. Phys. 71:129 (1979).
20. I. R. McDonald, Molec. Phys. 23:41 (1972).
21. D. J. Adams and I. R. McDonald, J. Phys. C 7:2761 (1974); 8:2198 (1975).
22. D. J. Adams and I. R. McDonald, Physica 79 B:159 (1975).
23. G. É. Norman and V. S. Filinov, High Temp. USSR 7:216 (1969).
24. D. J. Adams, Molec. Phys. 28:1241 (1974).
25. L. A. Rowley, D. Nicholson, and N. G. Parsonage, J. Comput. Phys. 17:401 (1975).
26. D. J. Adams, Molec. Phys. 32, 647 (1976); 37, 211 (1979).
27. J. E. Lane and T. H. Spurling, Aust. J. Chem. 29: 2103 (1976); 31: 465 and 933 (1978).
28. L. A. Rowley, D. Nicholson, and N. G. Parsonage, Mol. Phys. 32: 365 and 389 (1976); J. Comput. Phys. 26:66 (1978).
29. J. E. Lane, T. H. Spurling, B. C. Freasier, J. W. Perram, and E. R. Smith, Phys. Rev. A20: 2147 (1979).
30. W. van Megen and I. K. Snook, Molec. Phys. 39:1043 (1980).
31. G. M. Torrie and J. P. Valleau, Chem. Phys. Lett. 65:343 (1979).
32. J. A. Barker and D. Henderson, Rev. Mod. Phys. 48:587 (1976).
33. J. N. Cape and L. V. Woodcock, J. Chem. Phys. 72:976 (1980).
34. F. F. Abraham, Phys. Rev. Lett. 44: 463 (1980).
35. "Stochastic Molecular Dynamics", D. Ceperley and J. Tully, eds., NRCC, Lawrence Berkeley Laboratory (1979).
36. A. J. Stace, Molec. Phys. 38:155 (1979).
37. G. E. Murch and R. J. Thorn, Phil.Mag. 35:493; 36:517 and 529 (1977).

THE CALCULATION OF FREE ENERGIES USING COMPUTER SIMULATION

N. Quirke

Department of Chemistry
Royal Holloway College
Egham, Surrey TW20 OEX
England

INTRODUCTION

In the study of the thermodynamics of model systems there are two related questions which are of primary importance,

a) what is the equilibrium state of the system given certain constraints for example, that the pressure and temperature be constant, and

b) where are the phase transitions, if any, located with respect to the constraints?

In the first case we may be interested in knowing the equilibrium concentration of vacancies or defects in an otherwise perfect lattice or the most stable structure given a particular potential. In the second case we may wish to pinpoint the value of, for example, the pressure at which one equilibrium phase or structure of a system becomes unstable and changes to another. The answers to both these questions can be obtained from the particular thermodynamic potential appropriate to the imposed constraints, since this function is a minimum (or maximum) for the most stable state of the model system. Given that we wish to study the behaviour of a model system using conventional independent variables, either the Helmholtz free energy (A) or the Gibbs free energy (G) will be required. The Helmholtz free energy is the thermodynamic potential for conditions of constant numbers of particles (N), constant volume (V) and constant temperature (T), the canonical ensemble conditions. The Gibbs free energy is the thermodynamic potential at constant N,T and Pressure, the isothermal-isobaric ensemble conditions. The chemical potential (μ) can be obtained from G for a single component system by dividing by the number of particles, N. Using N,P,T as independent variables a typical first order phase transition (e.g. fluid-liquid) can be located by plotting the chemical potential against Pressure on both sides of the approximate transition point and extrapolating back until the two curves

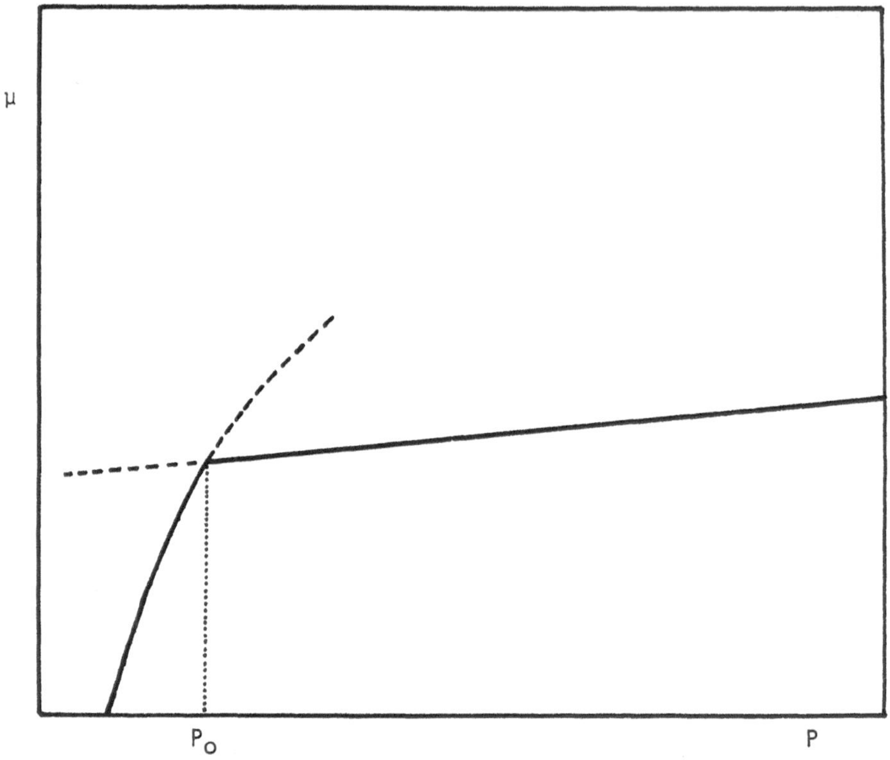

Fig. 2. Typical energy distribution functions for a molecular model[12] of liquid nitrogen. In this case Θ is the configurational potential energy for 32 two centre Lennard-Jones nitrogen molecules in a Monte Carlo simulation at a reduced density of $\rho\sigma^3 = 0.70$, corresponding to the triple point region.

intersect, as in Figure 1. In this lecture we shall be concerned with the application of the techniques of computer simulation to the problem of calculating these thermodynamic potentials for model systems.

Thermodynamic quantities are related to microscopic quantities i.e. positions and velocities, through statistical mechanics. Very often we are concerned with "Mechanical" thermodynamic properties such as energy and pressure which can be written as statistical mechanical averages of functions of positions and velocities. These averages can be written in two ways, either as an average over time or as an average over phase space which because the kinetic part of the problem factors out of the partition function, becomes a weighted average over all possible configurations of the system. Computer simulation, for the physicist and chemist, is a method of performing such averages numerically. The molecular dynamics[1] technique performs time averages by solving the equations of motion appropriate to the constraints and Metropolis Monte Carlo, as Dr. Adams has explained[2], is a way of efficiently performing the configurational averages, again appropriate to the particular constraints.

Unfortunately for most constraints of interest, including those mentioned above, the thermodynamic potential or free energy cannot be expressed as an average of a function of the microscopic variables and cannot therefore be calculated from a computer simulation in the same manner as the energy or pressure. Nevertheless, methods can be devised that enable free energies to be calculated from the results of computer simulations and the next section briefly reviews some aspects of current work in this field.

II. Methods

In this section we look briefly at various methods of obtaining information about the free energy of model systems using computer simulation. Most work has been done on liquids using the Metropolis Monte Carlo technique although this may change in the future (see section III).

A. Thermodynamic Integration

In principle we can always integrate a derivative of the free energy calculated by computer simulation over a series of states between a state at a known value of the free energy to the required state. For example, the equation below gives the excess Helmholtz free energy (A) in terms of the integral of the pressure (P) with respect to number density (ρ).

$$\frac{A^{ex}}{Nk_BT} = \int_0^{\rho} \frac{1}{\rho'} \left[\frac{P}{\rho' k_BT} - 1 \right] d\rho' \qquad (1)$$

In practice two difficulties are encountered,

i) the number of intermediate states may be large and, of themselves, uninteresting;

ii) if a phase transition exists between the reference and required state, special, possibly computationally costly procedures may be required to create artificially stable states in the co-existance region.

Hansen and Verlet[3] in their study of the phase diagram of a system interacting via the atomic Lennard-Jones potential, created van der Waals loops in the liquid-gas co-existance region by limiting the allowed density fluctuations within the central box of the simulation. This created a reversible path from the fluid to the liquid side of the phase transition along which equation 1 could be integrated. Thermodynamic integration has until recently been the standard method of obtaining free energies. In the rest of this section we shall examine new techiques which are an attempt to improve a) the efficiency and b) the accuracy of free energy calculations.

B. Umbrella Sampling (4)

Although the free energy cannot be written as an ensemble average in the same straightforward way as the energy for example, the free energy difference between two model systems (labelled 0 and 1) can be expressed as a simple average, as below,

$$\frac{A_1}{k_B T_1} - \frac{A_o}{k_B T_o} = - \ln \left\langle \exp\left(- \frac{U_1}{k_B T_1} + \frac{U_o}{k_B T_o}\right)\right\rangle_o \qquad (2)$$

where $\langle \ldots \rangle_o$ is an ensemble average at constant N,V,T for the system with potential U_o.

From this equation it appears that taking the ensemble average of the exponential of the difference between the two reduced potentials will give the required free energy difference. Unfortunately for practical reasons in most cases this is not the case (see subsection D, and below for exceptions). Umbrella sampling is a trick that forces equation 2 to give a useful answer. In the following, we shall consider this equation in some detail since the quantities to be introduced will also be of use in the discussion of the other methods in this section.

Equation 2 is a shorthand form of

$$\Delta\left(\frac{A}{kT}\right) = - \ln \frac{1}{Q_o} \int \exp\left(- \frac{U_1}{k_B T_1} + \frac{U_0}{k_B T_o}\right) \exp\left(- \frac{U_o}{k_B T_o}\right) d^N q \qquad (3)$$

here $Q_o = N!Q_o'$, Q_o' is the configurational partition function for the system with potential U_o. Defining

$\Theta = -\frac{U_1}{k_BT_1} + \frac{U_o}{k_BT_o}$, we rearrange the integration over configurational space so that all parts where $\Theta'(q^N) = \Theta'$, a constant, are grouped together, and integrate over all Θ ,

$$\Delta(\frac{A}{k_BT}) = -\ln\int_{-\infty}^{\infty} \exp(\Theta')\{\frac{1}{Q_o}\int\exp(-\frac{U_o}{k_BT_O})\delta(\Theta(q^N)-\Theta')d^Nq\}d\Theta' \quad (4)$$

At this point a normalised energy distribution function can be defined as (5),

$$f_{U_o}(\Theta') = \frac{1}{Q_o}\int \exp(-\frac{U_o}{k_BT_O})\delta(\Theta(q^N)-\Theta')d^Nq \quad (5)$$

this is the probability density for finding a state with $\Theta(q^N) = \Theta$ at constant N,V,T. This distribution is the basis of the Overlap Ratio Method (see method C). Equation 2 has now become,

$$\Delta(\frac{A}{k_BT}) = -\ln\int_{-\infty}^{\infty} f_{U_o}(\Theta)\exp(\Theta)\,d\Theta \quad (6)$$

In order to obtain $\Delta(A/k_BT)$ we must know $f_{U_o}(\Theta)$. An estimate of this type of function can be obtained from a metropolis Monte Carlo simulation at constant N,V,T. In such a simulation with potential U, the probability of finding a state with some function of the co-ordinates, $\theta(q^N)$ say, within the range $\theta\pm\delta\theta$ where $2\delta\theta$ is the unit of energy used to build the histogram, is

$$f'_u(\theta) = M_u(\theta)/m$$

where $M(\theta)$ is the number of equilibrium configurations for which $\theta(q^N) = \theta\pm\delta\theta$ and m is the total number of equilibrium configurations generated. The prime indicates that this quantity has been estimated from a computer simulation. For a sufficient number of equilibrium configurations of the system this becomes equal to a probability density of the type defined in equation 5,

$$f'_u(\theta) \rightarrow f_u(\theta).$$

In terms of the primed distribution equation 6 is now

$$\Delta(\frac{A}{k_BT})' = -\ln\int_{-\infty}^{\infty} f'_{u_o}(\Theta)\exp(\Theta)\,d\Theta \quad (7)$$

All is not well however, since by necessity m is a finite number and therefore f' will be zero when f is very small, for example $< 10^{-7}$. These are usually just the regions where the product,

$$f_{u_0}(\Theta)\exp(\Theta) \tag{8}$$

is large. It is now clear that the reason why equation 2 is true in principle but not in practice is that usually we do not obtain enough information on f from one metropolis Monte Carlo simulation. For the case where $U_0 = U_1$ and only the temperatures are different however, this equation is more useful. Singer and Singer[29] in their work on Lennard-Jones mixtures were able to calculate free energy differences between fluids whose temperatures differed by about 10% from an energy distribution obtained at one temperature only.

One way of extending the usefulness of equation 2 was suggested by Torrie and Valleau in their study of the spherical Lennard-Jones liquid[6]. They forced the Monte Carlo simulation into the interesting region of Θ by biasing the choice of configurations in an extra Monte Carlo run with potential U_0. By the proper choice of bias, a distribution $f'_B(\Theta)$ was obtained which covered the required range of Θ and was related to the original $f'_{U_0}(\Theta)$ through,

$$f'_{u_0}(\Theta) = \frac{f'_B(\Theta)/\omega(\Theta)}{\langle 1/\omega\rangle_B} \tag{9}$$

where $\omega(\Theta)$ is the biasing function and $\langle..\rangle_B$ is on average with respect to the biased choice of configurations. It can in fact be obtained from equation 9 in the region of Θ, where $f'_B(\Theta)$, the 'umbrella' distribution, overlaps with $f'_{U_0}(\Theta)$. This procedure allowed the evaluation of the product given in equation 8 in all regions of Θ for which it was significant.

A problem with this method is that there is no general prescription for $\omega(\Theta)$. However, using the $1/r^{12}$ part of the Lennard-Jones potential for the reference system, Torrie and Valleau[7] were able to reproduce the free energy data of Hansen and Verlet[3] and others for the Lennard-Jones liquid. They found very good agreement despite the fact that they used only 32 particles in the central box of the simulation. The implication being that free energy differences are relatively insensitive to the number of particles used.

C. The Overlap Ratio Method

The need to guess the weighting function required to create the umbrella distribution can often be avoided by calculating $f_u(\Theta)$ in both the reference and required system since for some potentials[9] the free energy difference between two systems with reduced potentials U_0/k_BT_0 and U_1/k_BT_1 can be written, at constant NVT, from equation 5 as,

$$\exp\left(\Delta\left(\frac{A}{k_B T}\right)\right) = \frac{fu_o(\Theta)}{fu_1(\Theta)} \exp(\Theta). \tag{10}$$

It is useful to note that we are not limited to the NVT (canonical) ensemble. At constant NPT we have for the Gibbs free energy (G),

$$\exp\left(\Delta\left(\frac{G}{kT}\right)\right) = \frac{fu_o(\Phi)}{fu_1(\Phi)} \exp(\Phi) \tag{11}$$

where $\Phi = -\frac{H_1}{k_B T_1} + \frac{H_o}{k_B T_o}$, H = Enthalpy, V = Volume.

These equations will be useful as long as distributions in each system are sufficiently large for similar values of Θ or Φ so that the primed (simulated) distributions overlap. In those cases where this is not the case intermediate distributions will be required which link both fu_o and fu_1. These can usually be obtained by calculating distributions at intermediate values of some system parameter (Multistage sampling[8]) or by designing artificial potentials which, when used in the simulation, provide a wide flat distribution connecting fu_o and fu_1,[9].

Although for some purposes, such as testing a perturbation theory a free energy difference is all that is required, often we will wish to convert the calculated difference into an absolute free energy. In order to do so, the free energy of one system must be known. For the liquid state a convenient set of reference fluids is provided by the hard potentials e.g. hard spheres, hard diatomics, since accurate analytic equations of state are available[10,11]. Jacucci and Quirke[9] used the hard diatomic fluid as a reference fluid for a two centre Lennard-Jones model of Nitrogen[12] obtaining free energies for the dense liquid. For the sub critical temperatures investigated all the nitrogen distributions (see figure 2) overlapped sufficiently to allow a straightforward application of equation 10. However, the distribution of the highest temperature required (112°K) did not overlap with that of the reference hard diatomic fluid (see reference 9). Instead of obtaining overlapping distributions at progressively higher temperatures until overlap was obtained with the hard diatomic distribution, an artificial potential was designed to bridge the gap in one extra simulation. All calculations were performed with 32 molecules which has the benefit not only of faster equilibration times but more importantly, much wider distributions. The free energies obtained in this way were in good agreement with a successful perturbation theory[13,14] and as can be seen in table 1, the pressures calculated from a fit to free energies obtained in more extensive calculations[15] are in excellent agreement with

a) the 32 particle pressures, calculated directly in the simulations,

Table 1.

$\rho\sigma^3$	$\frac{P\sigma^3}{\varepsilon}$			
	A	B	C	D
0.60	2.85	2.88		
0.62	3.86	3.61	3.77	~3.8
0.64	5.04	5.01	5.08	
0.66	6.42	6.15	6.63	
0.68	8.02	8.22	8.46	
0.70	9.84	9.62	10.63	~10.0

Pressures for a molecular model of liquid nitrogen[12].

A Obtained by differentiation of the fitted free energies obtained using the overlap ratio method with 32 particles[15].
B The measured pressure in the simulation used to calculate the free energy[15].
C Perturbation theory pressures[13,14].
D Extrapolated 500 particle pressures[13,14].

b) the perturbation theory pressures and,
c) the extrapolated pressures from much larger simulations.

The overlap ratio method is not restricted to the liquid state. For example, it has been used to calculate the equilibrium concentration of lattice vacancies in a rare gas crystal[16], using the equation 17

$$\frac{m}{N} = \exp(-\Delta G_V/k_BT)$$

where m is the number of vacancies, N the number of lattice sites and ΔG_V the Gibbs free energy associated with the formation of a vacancy at constant pressure.

D. The Virtual Overlap Method[9]

This method is an alternative to the overlap ratio procedure described in sub section D, when the reference potential is one of the hard sphere, hard diatomic class of potentials. It is based on the fact that the Boltzman factor of a hard potential is either zero or unity. Let U_0 be the two centre Lennard-Jones model (Nitrogen

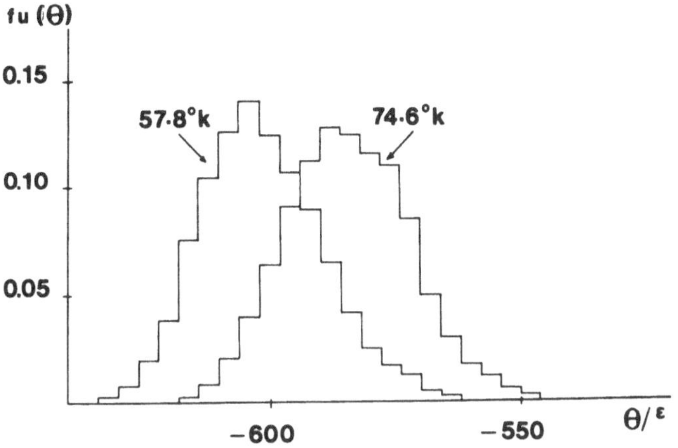

Fig. 2. Typical energy distribution functions for a molecular model[12] of liquid nitrogen. In this case Θ is a configurational potential energy for 32 two centre Lennard-Jones nitrogen molecules in a Monte Carlo simulation at a reduced density of $\rho\sigma^3 = 0.70$, corresponding to the triple point region.

potential(ϕ) referred to previously and U_1 the same potential but with a superimposed hard diatomic (h) potential with atomic diameter σ_h (see table 2). Then equation 2 becomes,

$$\frac{A_1}{k_B T_1} - \frac{A_o}{k_B T_O} = - \ln \langle \exp(\frac{-h}{k_B T_1}) \rangle_0 \tag{12}$$

The Boltzmann factor exp(-h) is either zero or one for all configurations generated with potential U_o depending upon whether any one site-site distance $R_{\alpha\beta}$ is smaller than σ_h. Therefore the average in equation 12 is just the fraction of the configurations generated for which all $R_{\alpha\beta} > \sigma_h$, which we shall call g. Note that in terms of energy distributions equation 12 is

Table 2.

U_1 †	U_o	Q_o/Q_1 (g)	N	Method	Label
$\phi+h(\sigma_h^*)$	$\phi+h(\sigma_h^*)$				
$\phi+h(0.92)$	ϕ	.03210	10^5	V.O.	A
$\phi+h(0.94)$	$\phi+h(0.92)$	.00871	10^5	V.O.	B
$\phi+h(0.95)$	$\phi+h(0.94)$	.02060	1.5×10^5	V.O.	C
$\phi+h(0.96)$	$\phi+h(0.95)$	.01580	10^5	V.O.	D
$\phi+h(0.9649)$	$\phi+h(0.96)$	.06170	10^5	V.O.	E
$\phi+h(0.9649)$	ϕ	5.615×10^{-9}		A*B*C *D*E	F
$h(0.9649)$	$\phi+h(0.9649)$	2.101×10^{-67}	2×10^5	O.R.	G
ϕ	$h(0.9649)$	1.180×10^{-75}		G*F	

$$\phi = 4\varepsilon \sum_{\alpha\beta} \left\{ \left(\frac{\sigma}{R_{\alpha\beta}}\right)^{12} - \left(\frac{\sigma}{R_{\alpha\beta}}\right)^{6} \right\}$$

$$h(\sigma_h) = \infty, \quad R_{\alpha\beta} < \sigma_h$$

$$= 0, \quad R_{\alpha\beta} > \sigma_h$$

† $\sigma_h^* = \sigma_h/\sigma$

Details of the calculation of the free energy difference between a molecular model[12] of liquid nitrogen at $T = 112^o$ K and $\rho\sigma^3 = 0.70$ and the hard diatomic fluid at the same density, where the hard diatomic has $\sigma_h = 0.9649\sigma$ [9].

ϕ = the nitrogen potential, $R_{\alpha\beta}$ is the distance between sites on the molecules.

U_o and U_1 are the potentials for each calculation, $Q_o/Q_1 = g$ is the calculated ratio of their respective configurational integrals (see text). N is the number of equilibrated configurations used to calculate g. V.O. refers to Virtual Overlap, O.R. refers to Overlap ratio method (sub-section C).

$$\Delta\left(\frac{A}{kT}\right) = -\ln \int_{-\infty}^{\infty} f_\phi(\phi-h)\exp\left(\frac{-h}{k_B T_1}\right)\, d(\phi-h) \qquad (13)$$

and that since only configurations with h = 0 can contribute we have

$$\Delta\left(\frac{A}{kT}\right) = -\ln \int_{-\infty}^{\infty} f_\phi^s(\phi)\, d\phi \qquad (14)$$

where s indicates that only configurations with no 'Virtual Overlaps' i.e. all $R_{\alpha\beta} > \sigma_h$ are counted. We see that, of necessity, the integrand has its maximum value at the same place as the calculated energy distribution function and as long as ϕ and h are not too different the free energy difference can be written and evaluated from

$$\frac{A_1}{k_B T_1} - \frac{A_0}{k_B T_0} = -\ln \frac{Q_0}{Q_1} = -\ln g \qquad (15)$$

Equation 15 is also true for the case $U_0 = \phi + h_1$, $U_1 = \phi + h_2$ where $\sigma_{h_2} > \sigma_{h_1}$.

It can be used in the following way. If the atomic diameter of the required reference hard diatomic potential is so large that g = 0 for a finite number of configurations, a chain of calculations, between systems with intermediate σ_h's, can be performed as shown in table 2. When finally we have the free energy difference between the Nitrogen state and the same state but with the required hard diatomic potential superimposed upon the Nitrogen molecules, then the free energy difference between this last state and the pure hard diatomic fluid is obtained using the overlap ratio method. This is a very easy calculation because the configurational behaviour of each system is dominated by the same hard core, therefore their energy distribution functions of ϕ overlap to a great extent.

Using 32 molecule systems this method predicted a configurational free energy difference between a state of Nitrogen at 112° K and reduced density of $\rho\sigma^3 = 0.70$ and a hard diatomic fluid with $\sigma_h = 0.9649\sigma$, of $\Delta A^c/NkT = -6.75$ once the long range corrections were added to the final result of table 2. Perturbation theory[13,14] gave $\Delta A^c/NkT = -6.78$ for the same state. Note that from table 2, the largest contribution to ΔA comes from the last step (labelled G). This is effectively the internal energy contribution to the free energy.

E. Bennett's Method[18]

Using statistical arguments Bennett derived the following

equations giving the most efficient method of calculating the free energy difference between two systems with reduced potentials $U_0' = U_0/kT_0$, $U_1' = U_1/kT_1$, from Monte Carlo simulations of finite length.

$$\frac{Q_0}{Q_1} = \frac{\langle I(U_0'-U_1' + c)\rangle_1}{\langle I(U_1'-U_0' - c)\rangle_0} \exp(+c) \tag{16}$$

$$I(x) = 1/1+\exp(x)$$

$$C = \ln\left(\frac{Q_0}{Q_1}\frac{n_0}{n_1}\right) \tag{17}$$

The fraction n_0/n_1 is the ratio of the number of independent configurations generated in each Monte Carlo simulation and is usually put equal to unity. From an initial guess for c, the required free energy difference is obtained by iterating equations 16 and 17 until convergence is achieved. This method gives an accurate answer under the same conditions as the overlap ratio method i.e. that there is sufficient overlap between the distributions of the difference in reduced potentials calculated in both systems. For the atomic and molecular Lennard-Jones liquids the two methods give answers to within 0.1% of each other. Bennetts method is easy to programme but at the cost of the detailed check on the calculation that the other methods described here allow.

F. Grand Canonical Ensemble Monte Carlo Calculations

Recently Metropolis Monte Carlo schemes for the constraints of constant μ,V,T, the grand canonical ensemble, have become available [19,20,21]. The method has been used to fix points on the μ,P diagram (see figure 1) by choosing μ and observing the mean pressure in the Monte Carlo run. These fixed points can then be used to obtain the unknown coefficients in an equation of state which may be available from other sources. Adams[22,23] has used this approach to treat the liquid-gas transition in the spherical Lennard-Jones liquid with some success. One difficulty is that the algorithms for inserting particles into the system runs into difficulty as the density increases because fewer gaps appear in the generated configurations. However, better algorithms are appearing[24] so that this method appears to be quite promising.

G. Ghost Particle Method (Widom's equation)

By rearranging the partition function for N+1 particles in a volume V at temperature T, Widom[25] obtained the following equation for the excess chemical potential μ^{ex} of an N particle system.

$$\mu^{ex} = -kT \ln\langle \frac{1}{V} \int \exp(-\frac{\psi}{k_B T})\, d\underset{\sim}{r} \rangle_N \tag{18}$$

ψ is the interaction energy between the N particles and an extra particle which does not affect the average <...>. In order to evaluate the three dimensional integral a "ghost" particle is introduced at various points in the configurations of N particles and the average value of the Boltzmann factor obtained for each configuration. This approach is very appealing because it does not require any reference state apart from the ideal gas. Further, as long as the system studied is large enough the average in equation 18 can be evaluated using Monte Carlo or molecular dynamics. It has been used successfully to treat hard spheres[26] and for two centre Lennard-Jones molecular liquids[27,28] at low to medium densities. The method appears to fail in the triple point region because $\exp(-\frac{\psi}{hT})$ is almost everywhere small. However the methods used to evaluate the three dimensional integral have been of the Simpsons rule type which are usually rather bad in three dimensions. We expect that more efficient three dimensional integration algorithms should lead to a substantial improvement in the results at higher densities.

III. Discussion

We have looked at a variety of methods for calculating free energies using computer simulation. They can be grouped into two classes,

a) those that require a link of some sort to be forged between a known reference system and the system for which the free energy is required, methods A,B,C,D,E of the last section, and

b) those that obtain the free energy directly, methods F and G.

Of the first class, method A is often more computationally expensive than the others. A large part of the problem is the effort involved in obtaining the initial link between one state of the required system in the general thermodynamic region of interest and the reference system. Once this has been done the extra effort involved in obtaining results for neighbouring states at the same density is not very great. The direct methods F and G offer a more economical approach but for the moment are not as versatile as the first class. For example, there appear to be difficulties with both methods at very high densities.

Most work to date has been done using Metropolis Monte Carlo simulation, probably because this approach offers such a wide range of possible constraints. However, recently new molecular dynamics schemes have been suggested[30] which allow these simulations to be performed at constant temperature or constant pressure or both. The use of such molecular dynamics programmes with, for example, an improved Ghost particle method promises to provide a very powerful tool indeed for the study of phase transitions and related problems using computer simulation.

References

1. See the lecture by S. de Leeuw, page of these proceedings.
2. D.J. Adams, page of these proceedings.
3. J.P. Hansen, L. Verlet, Phys.Rev. 184, (1969) 151.
4. J.P. Valleau, G.M. Torrie in Modern Theoretical Chemistry, ed. B.J. Berne, Vol.5, Chapter 5 (Plenum, 1977) and references therein.
5. I.R. McDonald, K. Singer, J.Chem.Phys., 50 (1969) 2308.
6. G.M. Torrie, J.P. Valleau, Chem.Phys. Letts., 28, (1974) 578.
7. G.M. Torrie, J.P. Valleau, J.Comp.Phys., 23 (1977) 187.
8. J.P. Valleau, D.M. Card, J.Chem.Phys., 57 (1972) 5457.
9. G. Jacucci, N. Quirke, Molec.Phys., 40 (1980) 1005.
10. N.F. Carnaham, K.E. Starling, J.Chem.Phys., 51 (1969) 635.
11. T. Boublik, I. Nezbeda, Cyem.Phys.Letts., 46 (1977) 315.
12. P.S.Y. Cheung, J.G. Powles, 30 (1975) 921.
13. F. Kohler, N. Quirke, J.W. Perram, J.Chem.Phys., 71 (1979) 4128.
14. J. Fischer, J.Chem.Phys., 72 (1980) 5371.
15. N. Quirke, G. Jacucci (unpublished work).
16. G. Jacucci, M. Ronchetti, Solid.State.Comm. 33 (1980) 35.
17. D.R. Squire, W.G. Hoover, J.Chem.Phys., 50 (1969) 701.
18. C.H. Bennett, J.Comp.Phys., 22 (1976) 245.
19. D.J. Adams, Molec.Phys., 29 (1975) 307.
20. L.A. Rowley, D. Nicholson, N.G. Parsonage, J.Comp.Phys. 17 (1975) 401.
21. G.E. Normann, U.S. Filinov, High.Temp. USSR, 7 (1969) 216.
22. D.J. Adams, Molec.Phys., 32 (1976) 647.
23. D.J. Adams, Molec.Phys., 37 (1979) 211.
24. M. Mezei, Molec.Phys., 40 (1980) 901.
25. B. Widom, J.Chem.Phys., 39 (1963) 2808.
26. D.J. Adams, Molec.Phys., 28 (1974) 1241.
27. S. Romano, K. Singer, Molec.Phys., 37 (1979) 1765.
28. J.G. Powles (1980). Accepted for publication in Molecular Physics.
29. J.V.L. Singer, K. Singer, Molec.Phys., 24 (1972) 357.
30. H.C. Anderson, J.Chem.Phys., 72 (1980) 2384.

MONTE CARLO STUDY OF SOLID ELECTROLYTES

Y. Hiwatari and A. Ueda*

Department of Physics, Faculty of Science, Kanazawa University, Kanazawa 920, Japan

*Department of Applied Mathematics and Physics, Faculty of Engineering, Kyoto University, Kyoto 606, Japan

ABSTRACT

A simple version of the soft-core model for simple liquids for superionic conductors is proposed. That is, the model potential is composed of only the soft-core and Coulomb potentials, Monte Carlo simulations for two solid electrolytes, α-AgI and CaF_2, are made using a fine-grained lattice for a 108-ion system and 324- or 96-ion systems respectively, at several temperatures. For both systems we have observed sublattice disordering over an appropriate temperature range. We have obtained the partial pair distribution functions, one body distribution of ions and mean square displacement of ions as a function of the Monte Carlo time-step. Our results compare satisfactorily with experiments and molecular dynamics calculations. It is found for CaF_2 that as the temperature increases the fraction of anions in the tetrahedral locations shows a drop in magnitude over a narrow temperature range, and that results obtained for the fluctuations of potential energies in the superionic phase have a rather large system-size-dependency.

1. INTRODUCTION

We propose a simple model system with ionic potentials consisting of only soft-core and Coulomb potentials. Our model may be considered as version of the soft-core model for superionic conductors, which has been successfully applied to simple liquids [1,2]. On the other hand, the recently proposed excluded-volume model [3] is nothing but a version of the hard-core model [4]. The present model will be useful to extract the leading term of the pair potentials relevant to the phenomena.

We carried out Monte Carlo calculations for two solid electrolytes, α-AgI and CaF_2, with our model potentials. In order to avoid the time-consuming calculation of the electrostatic energies using Ewald sums, we used a fine-grained lattice model constructed by successive binary divisions of the unit cell, and which approximates a corresponding continuum system very accurately [5]. This method enables us to tabulate, in advance, potential energies for all possible separations on the lattice, and hence considerably reduce the computation time for the calculation of the potential energy. This method of calculation is, therefore, particularly useful to study various properties of the model as a function of temperature.

2. THE MODEL

We consider a model system with the following pair potential $\phi_{ij}(r)$ between any ions

$$\phi_{ij}(r) = A_{ij}\left\{\frac{\sigma_i+\sigma_j}{r}\right\}^n + \frac{Z_iZ_je^2}{r} , \qquad (1)$$

where i,j describe the types of ions; A_{ij} the repulsive strength; σ_i,σ_j the ion radii; Z_i,Z_j the fraction of charge. This model is considerably simplified since both the polarization and dispersion energy terms are neglected in eq. (1). We have used the following values of the potential parameters.
α-AgI: $\sigma_+=0.63$ Å, $\sigma_-=2.20$ Å, $|Z_+|=|Z_-|=0.6$, $A_{ij}=A=1.23\times10^{-2}(e^2/\text{Å})$ and n=7. These are the same values as those of Vashishta and Rahman [6]. The lattice constant, a, is 5.08 Å. The calculation was made with a system of 54 cations and 54 anions initially forming a bcc structure (anions) with cations distributed over suitable tetrahedral sites.

In the case of CaF_2, we assume for a simplicity that $A_{ij}=A$ and $\sigma_+=\sigma_-=\sigma$. The latter assumption is made as one of the most preferable condition for which the sublattice disordering of anions is expected [7]. The lattice constant was a=5.90 Å, with σ_+ and σ_- taken so that both ions are in contact at nearest heighbor distance, i.e., $\sigma_+=\sigma_-=(\sqrt{3}/8)a=1.28$ Å. The strength parameter A was fitted to the Kim-Gordon potential [8] at the nearest neighbor distance; this yields for $A=1.07\times10^{-2}(e^2/\text{Å})$. We used $Z_+=2$, $Z_-=-1$ and n=12. The choice for the value of n is based on the consideration in ref. [7]. The Monte Carlo calculations were performed on a 324-ion system and a 96-ion system, starting from a fluorite lattice.

3. RESULTS FOR α-AgI

The Monte Carlo simulations were performed at the temperatures k_BT/A=0.290, 0.387, 0.484, 0.581, 0.629, 0.678 and 0.968. The pair distribution functions corresponding to unlike ion pairs ($g_{+-}(r)$) and like ion pairs ($g_{--}(r)$) are shown in Fig. 1. The function $g_{--}(r)$ clearly shows the change of structures between our lowest temperature

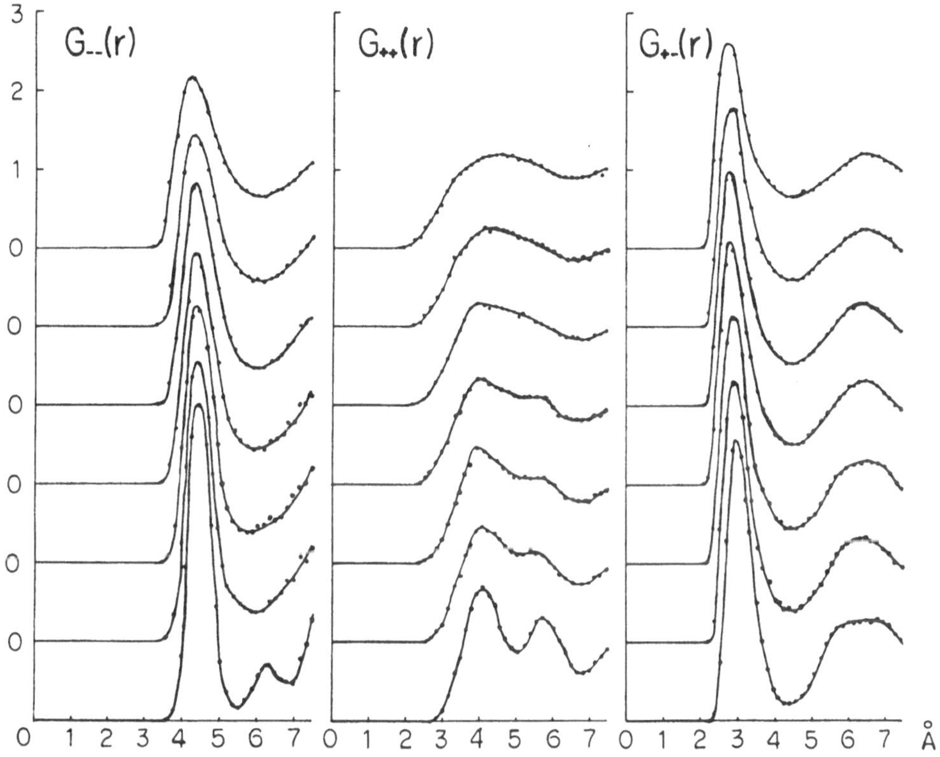

Fig. 1. Partial pair distribution functions $g_{--}(r)$ (anion-anion), $g_{++}(r)$ (cation-cation) and $g_{+-}(r)$ (cation-anion). From top to bottom the temperatures are 0.968, 0.678, 0.629, 0.581, 0.484, 0.387 and 0.290 (A/k_B), respectively. The second peak of the fcc structure appears for $g_{--}(r)$ at the lowest temperature.

and the others. The $g_{--}(r)$ at the lowest temperature displays a structure similar to that of fcc lattice, since the first two peaks occur at the nearest-neighbor spacing, r_0, and about $\sqrt{2}r_0$. On the other hand the function at higher temperatures presents a rather broad main peak as observed in all bcc structures. The main peak in the $g_{++}(r)$ is much weaker than that in the $g_{--}(r)$, and a clearly resolved second peak in the former at the lowest temperature is noted. The pair distribution functions for k_BT/A=0.484 show that the maxima of $g_{--}(r)$, $g_{++}(r)$ and $g_{+-}(r)$ occur at 4.3 Å, 4.0 Å and

2.8 Å with the heights 3.2, 1.5 and 3.1. These values are in good agreement with those of Vashishta and Rahman [6].

The mean square displacement generated from the MC process has revealed [8] that at $k_BT/A=0.290$ anions and cations are completely or almost non-diffusive. Only cations are diffusive at $k_BT/A=0.484$, and both ions become diffusive at $k_BT/A=0.678$.

From the one-body distribution of both ions, which are obtained by joining positions of ions averaged over 20 MC time, it has also been shown [8] that anions exhibit a stable regular arrangement at $k_BT/A=0.290$ and 0.484, but at $k_BT/A=0.678$ the arrangement becomes relatively disordered. At $k_BT/A=0.290$ cations appear to remain almost in localized regions, whereas cations at $k_BT/A=0.484$ are seen to move over a distance comparable with the nearest neighbor distance of the anion sublattice.

From these results and similar analyses for other temperatures it may be predicted that the superionic phase of the present model exists for $0.35 \lesssim k_BT/A \lesssim 0.65$; the phase change occurs around $k_BT/A \cong 0.35$ from the superionic to normal crystalline phase, accompanying the structural transition of the anion sublattice (bcc to fcc-like structure), and from the superionic to liquid phase around $k_BT/A \cong 0.65$. It follows that the temperature range of the superionic phase T_m(melting)/T_s(superionic transition) is predicted to be about 1.86, being very close to the experimental value (1.98) for α-AgI. The absolute temperature is, however, about 1.5 times higher than that of α-AgI. This discrepancy would be overcome if one uses a reduced value of A.

The time variations of x-, y- and z-coordinates of cations has shown [8] that cations spend an overwhelming time near tetrahedral sites (0,1/4,1/2), a rather alrge amplitude showing high anharmonicity ($k_BT/A=0.484$). This result supports the predictions on the cation density distribution for α-AgI by Hoshino et al. [9], Cava et al. [10], and Vashishta and Rahman [6].

4. RESULTS FOR CaF_2

Figure 2 shows the partial pair distribution functions $g_{ij}(r)$. The $g_{++}(r)$ clearly shows that cations form an fcc lattice. From low to high temperatures no structural change is observed. The $g_{--}(r)$ has much less clarity in its peak structure than that of $g_{++}(r)$. Compared with the Rahman's result [11], our pair distribution functions of the case (f) in Fig. 1 are found to agree quantitatively well with the former. In the former, the maxima of $g_{++}(r)$, $g_{--}(r)$ and $g_{+-}(r)$ occur at 4.0 Å, 2.9 Å and 2.2 Å with the heights 3.2, 1.9 and 4.2, respectively. In the present case the maxima occur at

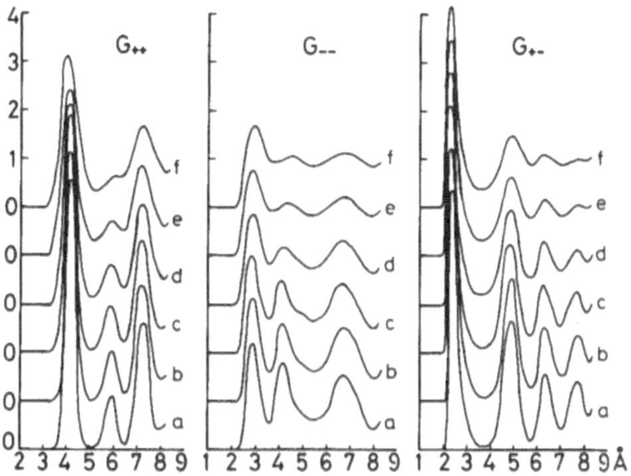

Fig. 2. Partial pair distribution functions for CaF_2. The temperatures are (a) 1.114, (b) 1.225, (c) 1.281, (d) 1.392, (e) 1.670 and (f) 1.949 (A/k_B), respectively.

4.0 Å, 2.9 Å and 2.3 Å with the heights 3.1, 1.7 and 4.2, respectively.

From the calculation of the mean square displacement it was found [12] that cations are non-diffusive for all temperatures we studied, while anions, in cases (d)-(f), are clearly diffusive. In cases (b) and (c), the anions are slightly diffusive, and in case (a) the anions are almost non-diffusive.

Figures 3(a),(b) show the potential energy per ion and fluctuations of the potential energies, respectively. In (a) data points seem to lie on a smooth continuous line, and no significant system-size dependence is observed. On the other hand, in (b) there is a rather large system-size dependence for the two high temperatures. From the result of 96-ion system a λ-like anomaly seems to occur near $k_BT/A \cong 1.3$, being similar to experimental specific heat of CaF_2 [13], while the result of 324-ion system shows a different behavior such that the specific heat increases as the temperature increases.

Figure 3(c) shows the fractional number of anions located at the tetrahedral and interstitial octahedral locations. It is found

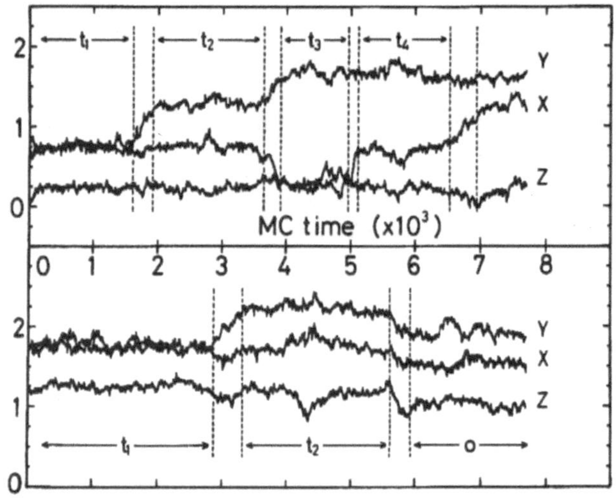

Fig.3.(a) Potential energy per ion in units of 14.39 Ev. (b) Potential energy fluctuations per ion: ▲ N=96 ion system, ● N=324. (c) Fraction of anions located at the tetrahedral (t) sites and octahedral (o) sites.

that there is a drop of the fraction of anions in t-sites locations around $k_BT/A \cong 1.3 \sim 1.4$, that is, for the low temperature side it is nearly one as expected, and for the high temperature side it decreases to about 0.7. On the other hand the fraction in the octahedral locations shows a smooth increase as the temperature increases. Near the transition the fraction in the octahedral locations

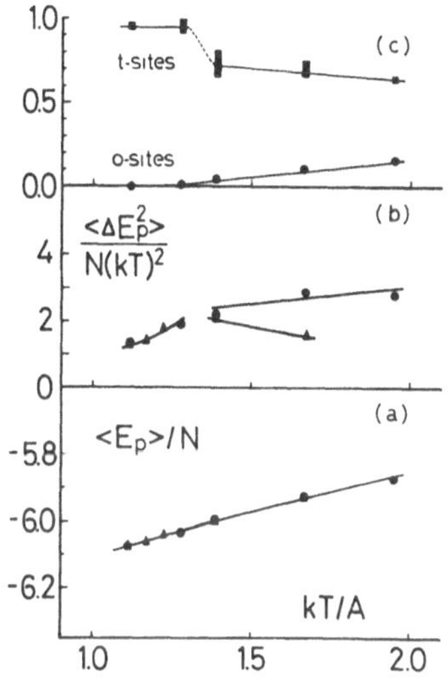

Fig. 4. Time variations of x-, y-, and z-coordinates of two arbitrarily selected anions for a MC run of $k_BT/A=1.670$. Unit length in the ordinate is the lattice constant of the fcc structure (5.90 Å). It is clear that anions undergo hopping through available tetrahedral sites designated by t_i or octahedral site designated by o.

reaches only a few percent, and therefore the transition can not be described by the usual two-site (tetrahedral and octahedral) model of the order-disorder type. This result is consistent with results of Dickens et.al. [14] and Dixon and Gillan [15], but is inconsistent with that of Axe et.al. [16].

Figure 4 shows the x-, y-, and z-coordinates of anions as a function of MC time. It is clear that anions oscillate about positions with rather large amplitudes, and that they travel between these positions in a very short period compared with the residence time of oscillations. In the upper figure, the anion moves between available tetrahedral locations t_i, that is, from $t_1 \to t_2$ in the (100) direction, from $t_2 \to t_3$ in the (110) direction and from $t_3 \to t_4$ in the (100) direction. In the lower figure the anion moves from $t_1 \to t_2$ in the (100) direction and from $t_2 \to o$ (octahedral location) in the (111) direction. The diffusive motion of anions occurs as discrete hops with a large ratio of residence to hopping times. This is consistent with the molecular dynamics results for CaF_2 [15] and for $SrCl_2$ [17]. Dixon and Gillan [15,17] has shown that more than 80% of hops occur in the (100) direction, about 10% go in the (110) direction, and only a few percent are in the (111) direction.

In the Monte Carlo calculation the time-step is not real time. Assuming that the MC time is proportional to the real time, i.e., $t_{MC}=\tau_s t$, we evaluated the ratio of the diffusion constant to τ_s. We obtained [12] its different temperature-dependence between the high and low temperature regions around $A/k_BT \cong 0.75$. This is similar to the same properties observed for the conductivities of superionic conductors of fluorites [18].

Finally we remark that the absolute temperature of the phase transition predicted from $k_BT/A \cong 1.3 \sim 1.4$ with $A=1.07\times10^{-2}(e^2/Å)$ is calculated to be $T_c=2300 \sim 2500K$, which is much higher than the experimental value for CaF_2, $T_c=1423K$. However this discrepancy would be partly overcome by using a reduced value of A. On the other hand our result shows that the temperature range of the superionic phase predicts a value greater than 1.4, which is also larger than that of $CaF_2 (\cong 1.2)$. Such a large temperature range of the superionic phase suggests that the present model using N=12 will be more suitable for PbF_2- or BaF_2- fluorites rather than CaF_2 as in the two first cases, $T_m/T_c=1.5$ and 1.3, respectively. It would thus be of interest to study the n-dependence of our model, as this should be responsible to describe the differences between superionic conductors of fluorites.

REFERENCES

1. Y. Hiwatari and H. Matsuda, Prog. Theor. Phys. 47 (1972) 741; H. Matsuda and Y. Hiwatari, Cooperative Phenomena, ed. H. Haken and M. Wagner (Springer-Verlag, Berlin, 1973) p. 250.
2. Y. Hiwatari, H. Matsuda, T. Ogawa, N. Ogita and A. Ueda, Prog. Theor. Phys. 52 (1974) 1105.
3. J.B. Boyce, T.M. Hayes, W. Stutius and J.C. Mikkelsen, Jr., Phys. Rev. Lett. 38 (1977) 1362; T.M. Hayes, J.B. Boyce and J.L. Beeby, J. Phys. C11 (1978) 2931.

4. B.J. Alder and W.G. Hoover, Physics of Simple Liquids, ed., H.N.V. Temperley, J.S. Rowlinson and G.S. Rushbrooke (North-Holland, Amsterdam, 1968) Chap. 4.
5. T. Ichimura, T. Shiotani and A. Ueda, Prog. Theor. Phys. 60 (1978) 941.
6. P. Vashishta and A. Rahman, Phys. Rev. Lett. 40 (1978) 1337.
7. Y. Hiwatari, Phys. Lett. A 75 (1980) 426.
8. Y. Hiwatari and A. Ueda, J. Phys. Soc. Jpn. 48 (1980) 766.
9. S. Hoshino, T. Sakuma and Y. Fujii, Solid State Commun. 22 (1977) 763.
10. R.J. Cava, F. Reidinger and B.J. Wuensch, Solid State Commun. 24 (1977) 411.
11. A. Rahman, J. Chem. Phys. 65 (1976) 4845.
12. Y. Hiwatari and A. Ueda, J. Phys. Soc. Jpn. 49 no. 6 (1980).
13. M. O'Kneeffe and B.G. Hyde, Philos. Mag. 33 (1976) 219.
14. M.H. Dickens, W. Hayes and M.T. Hutchings, J. Physique Coll. C7, 37, Supp. 12 (1976) 353.
15. M. Dixon and M.J. Gillan, J. Phys. C 11 (1978) L165.
16. J.D. Axe, S.M. Shapiro and N. Wakabayashi, Proc. Int. Conf. on Fast Ion Conductors, ed. G.D. Mahan and W.L. Roth (Plenum Press, New York, 1976).
17. M. Dixon and M.J. Gillan, Proc. Int. Conf. on Fast Ion Transport in Solid, Electrodes and Electrolytes, ed. P. Vashishta, J.N. Mundy and G.K. Shwnoy (North-Holland, Amsterdam, 1979) p. 701.
18. For example, see J.B. Boyce and B.A. Huberman, Phys. Rep. 51 (1979) 189.

MOLECULAR DYNAMICS WITH CONSTRAINTS

H.J.C. Berendsen and W.F. van Gunsteren

Laboratory of Physical Chemistry
University of Groningen
Nijenborgh 16
9747 AG Groningen
The Netherlands

A method is described to include constraints such as bond lengths and bond angles into molecular dynamics simulations, while retaining cartesian coordinates. The effects of constraints on physical behaviour and computational efficiency are evaluated. A Fortran subroutine (SHAKE) is appended.

1. INTRODUCTION

The modeling of large molecular systems by the method of molecular dynamics (MD) requires simulation of a large number of degrees of freedom, often over a considerable length of time. Practical limitations of computational effort form a strong incentive to reduce the number of degrees of freedom as much as possible while retaining a truthful simulation of the physical characteristics of interest. In principle there are two ways to achieve this reduction:

a. Less relevant degrees of freedom can be ignored and their influence approximated by a combination of mean force interactions, stochastic forces and frictional forces. This approach is treated elsewhere in this volume (W.F. van Gunsteren and H.J.C. Berendsen, Stochastic Dynamics).

b. Hard degrees of freedom, i.e., those corresponding to high-frequency normal modes, can be treated as if they were completely constrained and thus be eliminated from the system. This will increase computational efficiency by allowing a larger time step in the simulation. It is this approach that we will now consider in more detail.

Constraining degrees of freedom produces a dynamics that approximates the physical behaviour of the system. For this approximation to be valid in the sense that the motion of the system on a longer time scale is not greatly affected, it is required that the eliminated degrees of freedom are not strongly coupled to other degrees of freedom and thus their motion is separable from other modes. This is generally true for modes that are in a distinctly higher frequency range than all other modes. In molecules, likely candidates are bond lengths and bond angles, covering the near-infrared region of vibrational frequencies. In the case of bond angle vibrations, which are strongly mutually coupled, the frequency range extends into the far infrared and one may expect considerable overlap with and coupling to the "essential" degrees of freedom. Thus the applicability of constraints is more questionable in the case of bond angles than it is in the case of bond lengths.

In Section 2 we will review the general method, which was originally described by Ryckaert, Ciccotti and Berendsen [1], followed by a description of the iterative procedure SHAKE. A Fortran listing of SHAKE is given in the Appendix. In Section 3 we evaluate both the computational efficiency and the effects on the physical behaviour of a macromolecular system. The influence of the metric tensor of the hypersurface to which the motion is constrained on the physical properties is also considered in Section 3. Finally, the conclusions (Section 4) can be summarized as follows:

In (macro)molecular systems the conservation of bond length constraints is computationally efficient and has little influence on the physical properties of the system. Constraining bond angles, however, is not recommended while it distorts the physical behaviour of the system and yields no computational advantages.

2. MOLECULAR DYNAMICS WITH CONSTRAINTS

2.1 Classification of constraints

Constraints can be classified in the following categories [2]:

a. holonomic constraints are of the form

$$\sigma_k(\underset{\sim}{r}_1,\underset{\sim}{r}_2,\ldots,\underset{\sim}{r}_N,t) = 0 \qquad k = 1,\ell \tag{2.1}$$

for a system of N particles satisfying ℓ independent constraints. If the functions σ_k do not contain the time t explicitly, the constraints are called scleronomous; if σ_k are explicit functions of time (such as a particle that is constrained to a moving surface), the constraints are rheonomous.

b. nonholonomic constraints are any constraints that cannot be written in the form of eq. 2.1, e.g. a particle that is constrained to move above a given surface.

We shall restrict ourselves to scleronomous holonomic constraints, appropriate for time-independent bond lengths and bond angles. In practice the functions σ_k can be expressed in terms of distance between two particles

$$\underset{\sim}{r}_{ij}^2 - d_{ij}^2 = 0 \tag{2.2}$$

where

$$\underset{\sim}{r}_{ij} \equiv \underset{\sim}{r}_i - \underset{\sim}{r}_j \quad . \tag{2.3}$$

Also angle constraints can be satisfied by a distance criterium, as is shown in fig. 1, where the angle α is constrained by $\underset{\sim}{r}_{12}^2 - d^2 = 0$.

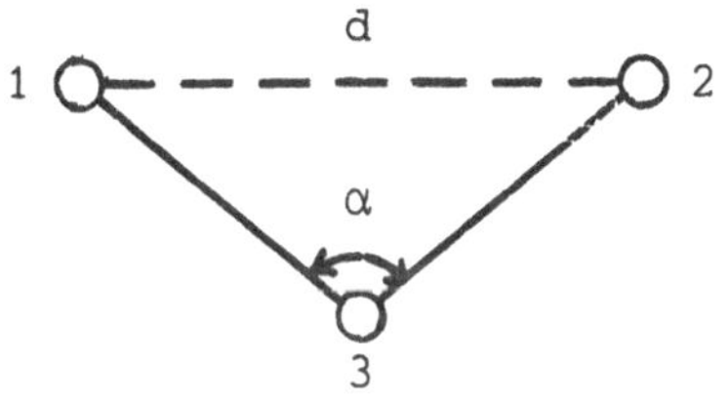

Figure 1.

This is strictly true only when the lengths of bonds that form the angle are constrained as well. In practice this is always the case, because one never considers bond angle constraints only.

2.2 Generalized coordinates

The classical way to conserve constraints is to use a set of generalized coordinates $\underset{\sim}{q}$, in which the constraints are separable. A system of N particles with ℓ constraints is described by a set of $3N-\ell$ generalized coordinates q_j, $j = 1,\ldots,3N-\ell$

$$\underset{\sim}{r}_i = \underset{\sim}{r}_i(q_1,\ldots,q_{3N-\ell}) \tag{2.4}$$

Equations of motion can be derived from Lagrange's equations, as follows. First write the kinetic energy T and the potential energy V as functions of $\underset{\sim}{q}$ and $\dot{\underset{\sim}{q}}$. The $3N-\ell$ equations of motion are then given by

$$\frac{d}{dt}\left(\frac{\partial L}{\partial \dot{q}_j}\right) - \frac{\partial L}{\partial q_j} = 0 \tag{2.5}$$

where L=T-V is the Lagrangian of the system. For a rigid body ($3N-\ell=6$) this procedure leads to the Newton-Euler set of equations for the transitional and rotational degrees of freedom. If internal degrees of freedom are involved the procedure is still straightforward, but leads to complicated sets of equations that cannot be handled for a molecule of only moderate complexity [1].

Thus, the method of generalized coordinates is not suitable for general MD algorithms.

Another formalism for treating the dynamics of rigid bodies linked by hinges has been described by Wittenberg [3] and used by Pear and Weiner [4] on the simulation of a four-atom chain. The formalism is extendable to long chains but is vastly more complicated than the method we describe below. Fixman [5] uses a method which is essentially equivalent to the matrix method for solving constraint equations [1].

2.3 Cartesian coordinates

An alternative solution [1] is obtained by solving Newton's 3N equations of motion while observing ℓ constraints of the type

$$\sigma_k(\underset{\sim}{r}_1,\ldots,\underset{\sim}{r}_N) = 0 \qquad k = 1,\ldots,\ell \tag{2.6}$$

by Lagrange's method of undetermined multipliers:

$$m_i\ddot{\underset{\sim}{r}}_i = -\nabla_i(V + \sum_{k=1}^{\ell} \lambda_k\sigma_k) \tag{2.7}$$

Here λ_k are the Lagrangian multipliers. In this way 3N equations of motion and ℓ equations of constraint (2.6) can be solved for $3N+\ell$ variables (r_i and λ_k). The procedure is equivalent to the addition of forces of constraint $\underset{\sim}{G}_i$ to every particle involved in a constraint:

$$m_i\ddot{\underset{\sim}{r}}_i = \underset{\sim}{F}_i + \underset{\sim}{G}_i \qquad \text{(total force)} \tag{2.8}$$

$$\underset{\sim}{F}_i = -\nabla_i V \qquad \text{(unconstrained force)} \tag{2.9}$$

$$\underset{\sim}{G}_i = \sum_{k=1}^{\ell} \lambda_k\nabla_i\sigma_k \qquad \text{(constraint force)} \tag{2.10}$$

where λ_k are determined such that the ℓ constraints (2.6) are satisfied at all times.

We shall now consider a particular algorithm, viz. that introduced by Verlet [6] which eliminates velocities and solves for coordinates up to fourth order in Δt . As was shown in [1], the method to solve constraints remains valid for algorithms of any order in Δt. We will return to precautions for different types of algorithms at the end of this section.

In the Verlet algorithm including constraint forces, the following step is made

$$\mathbf{r}_i(t+\Delta t) = 2\mathbf{r}_i(t)-\mathbf{r}_i(t-\Delta t) + \frac{\mathbf{F}_i(t)}{m_i}(\Delta t)^2 + \frac{\mathbf{G}_i(t)}{m_i}(\Delta t)^2 \tag{2.11}$$

or

$$\mathbf{r}_i(t+\Delta t) = \mathbf{r}'_i + \delta\mathbf{r}_i \tag{2.12}$$

where $\mathbf{r}_i(t+\Delta t)$ are the coordinates after a constrained step, $\mathbf{r}'_i$ are the coordinates after a normal MD step disregarding all constraints, and $\mathbf{r}_i$ are the corrections to be made as a result of the constraints. The conditions (2.6) are:

$$\sigma_k(\mathbf{r}_1(t),\ldots,\mathbf{r}_N(t)) = 0 \tag{2.13}$$

while, according to 2.10 and 2.11, the corrections can be written as

$$\delta\mathbf{r}_i = -\frac{(\Delta t)^2}{m_i}\sum_{k=1}^{\ell} \lambda_k \nabla_i \sigma_k(\mathbf{r}_1(t),\ldots,\mathbf{r}_N(t)) \quad . \tag{2.14}$$

If the k-th constraint concerns particles i and j:

$$\sigma_k = \mathbf{r}^2_{ij}(t) - d^2_{ij} \tag{2.15}$$

the derivatives are given by

$$\nabla_i\sigma_k = -\nabla_j\sigma_k = 2\mathbf{r}_{ij}(t) \tag{2.16}$$

Hence (2.14) becomes

$$\delta\mathbf{r}_i = -\frac{2(\Delta t)^2}{m_i}\Sigma_k \lambda_k\mathbf{r}_{ij}(t) \tag{2.17}$$

where the sum extends over all constraints involving the i-th particle. Eq. 2.17 means that corrections due to the constraint between particle i and j must be applied in the direction of the vector $\mathbf{r}_{ij}(t)$. Corrections to $\mathbf{r}'_i$ and $\mathbf{r}'_j$ are in opposite directions and are weighted by the inverse masses of i and j.

Consider for simplicity the two-particle case (fig. 2). There is only one multiplier λ. The correction for particle 1 is

$$\delta\mathbf{r}_1 = -\frac{2(\Delta t)^2\lambda}{m_1}\mathbf{r}_{12} = \frac{g}{m_1}\mathbf{r}_{12} \tag{2.18}$$

and for particle 2

$$\delta\mathbf{r}_2 = \frac{2(\Delta t)^2\lambda}{m_2}\mathbf{r}_{12} = -\frac{g}{m_1}\mathbf{r}_{12} \tag{2.19}$$

Here g is an unknown parameter, to be solved from the constraint condition

$$\{\mathbf{r}_1' + \frac{g}{m_1}\mathbf{r}_{12} - (\mathbf{r}_2' - \frac{g}{m_2}\mathbf{r}_{12})\}^2 = d^2 \tag{2.20}$$

or

$$2g\left(\frac{1}{m_1} + \frac{1}{m_2}\right)\mathbf{r}_{12}'\cdot\mathbf{r}_{12} + g^2\left(\frac{1}{m_1} + \frac{1}{m_2}\right)^2 \mathbf{r}_{12}^2 = d^2 - \mathbf{r}_{12}'^2 \tag{2.21}$$

In this case there is one equation of the second degree in g which can be solved exactly. In the general case there are ℓ equations of the type 2.21, with ℓ unknown parameters g_k, each equation generally involving several g_k's. Thus we have a system of ℓ quadratic equations. When the equations are linearized (i.e., terms in g_k^2 are neglected), a set of ℓ linear equations is obtained that can be solved by matrix inversion. The first-order corrections thus obtained are then applied and the procedure is iterated until all constraints are satisfied to within a specified tolerance. This procedure was used successfully for n-butane [1,7] and n-decane [7]. For butane

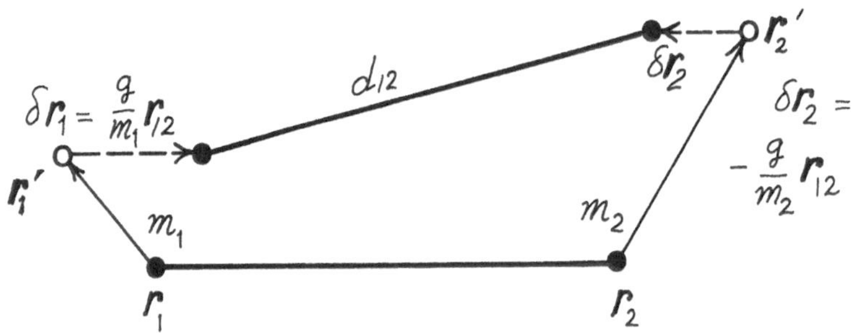

Fig. 2. Coordinate resetting for a diatomic molecule.

the procedure reduces computer time per step by a factor of 2.5 relative to the use of generalized coordinates.

2.4 The procedure SHAKE

An alternative way to solve the ℓ non-linear equations is to solve eq. 2.21 to first order for each constraint, treating all constraints in succession, and then iterating the procedure until all constraints are satisfied to within a specified tolerance. This is accomplished by a subroutine called SHAKE, the Fortran listing of which is given in the Appendix, together with indications for its use. The implementation of SHAKE in a MD algorithm is considerably more simple than the matrix method is, particularly for large molecules. Both methods yield the same numerical results within their specified tolerances. Comparison between both methods applied to a decane molecule has shown that computer time for SHAKE depends strongly on the required accuracy (time doubles when the relative accuracy is increased from 10^{-6} to 10^{-8}), while the matrix inversion method obtains a higher accuracy at little additional expense. For accuracies lower than 10^{-7} SHAKE is faster.

In practical computation one is confronted with the question how to set the tolerance for SHAKE. This will depend on the accuracy required for the MD simulation itself. A useful criterion for the accuracy of the MD algorithm is the rms fluctuation of the total energy (provided one uses a microcanonical non-stochastic simulation), expressed as a fraction of the rms fluctuation of the kinetic energy. Reliable simulations can be expected for relative total energy fluctuations of several %, although even higher values still seem to produce reliable statistics [8]. The best procedure for setting the tolerance <u>tol</u> of SHAKE is to determine by trial runs with increasing <u>tol</u> at what value of <u>tol</u> the total energy fluctuations begin to increase. A definite setting of <u>tol</u> 10x below this value is recommended. In practice a value of 10^{-5} is often appropriate.

It is also possible to estimate an upper limit for the noise in the total energy due to SHAKE if one has knowledge about the magnitude of force fluctuations $\Sigma_i \langle \underset{\sim}{F}_i^2 \rangle$ in the system studied. These force fluctuations can be estimated from the temperature and the largest force constants present in the system. Assume that SHAKE produces random errors in the particle positions $\Delta \underset{\sim}{r}_i$ which are neither correlated to each other nor to the forces $\underset{\sim}{F}_i$. This positional noise produces fluctuations in the energy:

$$\langle (\Delta E)^2 \rangle = \langle (\sum_i \underset{\sim}{F}_i \cdot \Delta \underset{\sim}{r}_i)^2 \rangle \approx \langle (\Delta \underset{\sim}{r}_i)^2 \rangle \sum_i \langle \underset{\sim}{F}_i^2 \rangle \tag{2.22}$$

or

$$\sqrt{\langle (\Delta E)^2 \rangle} < \delta \sqrt{\sum_i \langle \underset{\sim}{F}_i^2 \rangle} \tag{2.23}$$

where δ is the absolute tolerance of SHAKE (δ=<u>tol</u>* bond length).

2.5 Application of SHAKE to other MD algorithms

Thus far the introduction of a correction step to correct for constraints has been demonstrated only for the Verlet algorithm. It is equally possible to introduce this correction into other algorithms, although care has to be taken that instabilities are avoided that may result from the propagation of errors.

The general class of predictor-corrector algorithms has been treated and evaluated extensively in ref. [9], where procedures to incorporate SHAKE have been detailed. It appears necessary to incorporate two SHAKE corrections per step in predictor-corrector ("Gear") algorithms. Such algorithms, if used with constraints, are limited to representations in which the history of the trajactory, used for prediction, is represented by forces at previous times.

Higher-order predictor-corrector algorithms appear useful in those cases where the forces are predictable to a reasonable extent. Such cases are for example systems with harmonic forces which are not heavily damped, such as macromolecules and solid state systems. When hard harmonic degrees of freedom are constrained, predictability of the forces is generally reduced to such an extent that higher-order algorithms are not of great advantage. Thus, for constraint dynamics it generally suffices to use simple schemes such as the Verlet algorithm.

3. COMPUTATIONAL AND PHYSICAL CONSEQUENCES OF CONSTRAINTS

The computational efficiency as well as the influence on physical behaviour of bond length and bond angle constraints have been tested for macromolecular dynamics [9,10] and for energy minimization procedures [11]. The tests were made on an isolated protein molecule bovine pancreatic trypsin inhibitor (BPTI), built up from a single macromolecular chain of 58 aminoacids containing 454 "united atoms" (i.e. hydrogen atoms are not considered separately but are contracted into the heavy atom to which they are bound) and 4 internal water molecules. Thus the molecule possesses 1374 cartesian degrees of freedom; there are 468 possible bond length constraints and 626 possible bond angle constraints. The remaining degrees of freedom are 274 dihedral angles (hindered rotation about bonds, both in the backbone and in the side chains of the protein) and six trivial translational and rotational degrees of freedom. The potential energy functions used include bond stretching, bond angle bending, dihedral angle twisting, hydrogen bonds and non-bonded and electrostatic interactions. The main results are summarized below. In view of the conclusions of Section 2.5 we will restrict our discussion to the Verlet algorithm. For predictor-corrector algorithms we refer to ref. [9]. In all cases constraint resetting was performed by the procedure SHAKE.

3.1 Computational efficiency

When constraints are included, computer time per step increases due to the incorporation of SHAKE. For bond constraints CPU time increased by a factor 1.3 (from 1.2 to 1.6 s on a Cyber 74-16). For bond angle constraints SHAKE took considerably more time. A gain in efficiency therefore can result only from a considerably larger time step. The incorporation of bond length constraints allowed a time step 4x larger than without constraints for the same accuracy (fluctuation in total energy). Thus a gain in computational efficiency by a factor 3 was obtained. When bond angles were constrained in addition to bond lengths, no further increase in the time step was obtained. This is apparently due to the overlap of bond angle vibrational frequencies with other vibrational modes of the molecule. This overlap is also apparent from the analysis by McCammon et al. [12] of a MD simulation on BPTI.

Similar conclusions on the use of constraints were reached for energy minimization procedures [11]. It is well possible to build constraints into energy minimization procedures by using SHAKE, preferably in conjunction with conjugate gradient methods. In the case of bond length constraints SHAKE allows at least a three times larger step size than when no constraints are applied, resulting in a faster sampling of configuration space. But, when applying bond-length and bond-angle constraints, the step size is reduced by a factor of 3, compared to a non-constrained energy minimization in order to let SHAKE converge.

3.2 Influence on physical behaviour

The fact that bond length constraints allow an appreciable increase in time step while angle constraints do not, is in itself an indication that bond stretching vibrations are separable and constraining them is physically allowed, while bond angle deformation modes cannot be separated and constraining them is physically unwarranted. This conclusion, however, should be tested more directly by comparing results of constrained and non-constrained MD runs.

The obvious test on the equivalence of different MD methods is a comparison of individual trajectories. A requirement of detailed reproducibility of trajectories, however, is far too stringent because any computational error, however small, causes an eventually diverging deviation of the trajectory. Instead, the question to be asked is whether statistical ensemble averages (time-independent as well as time-dependent) are reproduced correctly. This requires comparisons based on full-scale MD runs. On a limited time scale such comparisons were made by comparing MD runs on BPTI during 5 ps. [10].

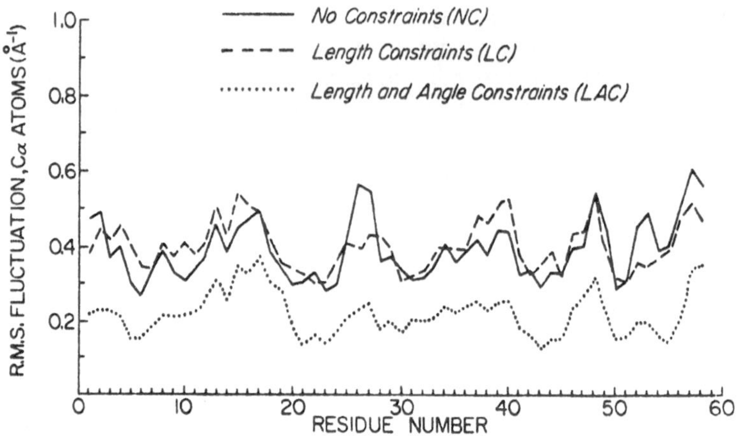

Fig. 3. Positional fluctuation in the protein BPTI.

In fig. 3, rms fluctuations of the positions of all 58 backbone α-carbon atoms are plotted for three cases: no constraints, bond length constraints, and bond length and angle constraints. It is observed that bond length constraints reproduce the structural fluctuations fairly well; agreement can be considered to be within statistical accuracy. Bond angle constraints, on the other hand, reduce structural fluctuations considerably. Although general features of the fluctuations are still qualitatively reproduced, the reduction in magnitude by a factor of two indicates that a significant part of configuration space is made inaccessible by constraining bond angles. We consider this to indicate that constraining bond angles significantly affects statistical ensemble averages.

Also for minimization procedures it was concluded [11] that the use of bond angle constraints produce significantly different results. Bond length constraints have little influence on the minimized configuration that is finally obtained.

Bond length constraints not only conserve time-independent ensemble averages, but also time-dependent properties. An example of the latter is shown in fig. 4, where spectral densities of the motion of a bond angle and a dihedral angle are compared for MD simulations with and without bond length constraints. The main features of the spectral distribution are present in both simulations. There is no evidence for coupling to vibrational bond stretching modes.

3.4 Influence of the metric tensor

When constraints are conserved in a molecular dynamics simulation or in any calculation of statistical mechanical ensemble averages, a rather subtle problem arises. A constrained system is described by a hamiltonian in generalized coordinates and momenta not including those

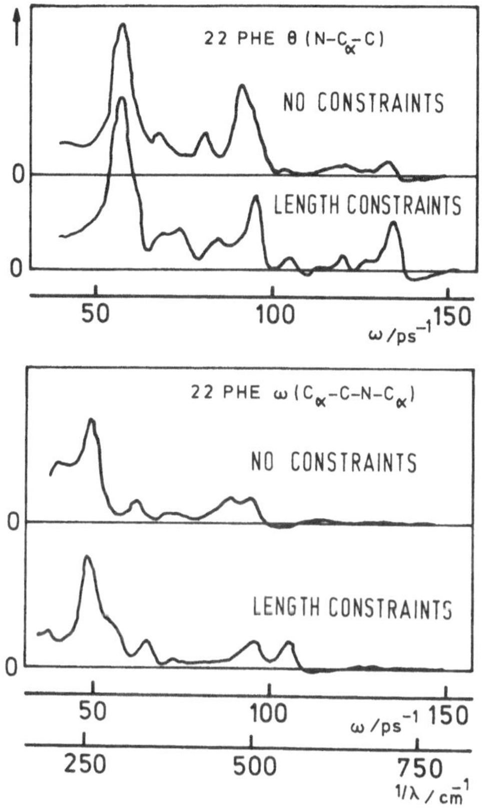
22 PHE θ (N-Cα-C)
NO CONSTRAINTS
LENGTH CONSTRAINTS
0
0
50
100
150
ω/ps⁻¹
22 PHE ω (Cα-C-N-Cα)
NO CONSTRAINTS
LENGTH CONSTRAINTS
0
0
50
100
150
ω/ps⁻¹
250
500
750
1/λ / cm⁻¹

Figure 4.

of the constrained degrees of freedom. This is not equivalent to a hamiltonian that is obtained as a limiting case of a hamiltonian of full dimensionality, in which the corresponding degrees of freedom are considered as harmonic oscillators in the limit of infinite force constants. In the latter case the momenta conjugate to the hard degrees of freedom do not vanish, as they do in the constrained case. While at first sight this may seem a trivial difference resolved by proper normalisation, closer inspection reveals that distribution functions in configuration space may differ in the two cases. This is particularly so for systems with internal degrees of freedom, where constrained degrees of freedom are not orthogonal to non-constrained ones. Integration over the hard degrees of freedom results in a factor depending on the soft variables, while such a factor is nonexistent in the constrained case. This produces a weighting factor in distribution functions for equilibrium properties on the constrained hypersurface. The weighting factor is determined by the metric properties of the coordinate transformation and can be expressed in terms of the determinant of the "mass-metric tensor" (see below) of the hypersurface to which the system is constrained.

In addition to this metric factor, another term in the equilibrium distribution function may result from dependence of the force constants of hard modes on soft variables. Such terms differ for the classical and quantum limits of the harmonic oscillator model and would cause considerable complications. However, in all usual force field models, harmonic oscillator force constants are functions of the hard variables only. Since this problem is thus neglected even in non-constrained cases, it is not directly related to the specific influence of constraints and we shall not consider it here.

The metric tensor problem has been, and still is, subject of considerable discussion and confusion in the literature. Gō and Scheraga [13] have considered consequences for static equilibrium properties of polymers and also reviewed previous discussions. Fixman [14] derived a powerful theorem that makes practical computation of the required correction term feasible. Fixman [15] also proposed that constrained dynamics simulations may be corrected by the addition of an extra potential energy term. This provides for correct simulations of static equilibrium averages, but does not guarantee correct dynamics. Pear and Weiner [4] and Helfand [15] have considered the influence on dynamics in more detail. Helfand [15] also gives a lucid example of a simple two-dimensional model illustrating some of the principles involved. Gottlieb and Bird [16] simulated a three-atom molecule in a box with spherical solvent molecules constraining bond lengths. Pear and Weiner [4] performed a Brownian simulation on a four-atom molecule (with 90° bond angles), constraining bond lengths and angles. Both simulations show clearly that the remaining internal degree of freedom is unevenly distributed if no corrections are applied, while Fixman's proposal for the correction seems quite appropriate. For butane the metric tensor correction in the potential has been compared to a mean force solvent potential by Chandler and Berne [17]. Finally van Gunsteren [18] has performed a stochastic simulation (using SHAKE) on butane without internal dihedral potential. In this simulation the influence of bond length constraints only was compared with that of both bond length and bond angle constraints.

We will now briefly review the theory leading to a correction term in the potential to be used in constraint dynamics, and discuss its consequences.

Consider a transformation from 3N cartesian coordinates $\underset{\sim}{x}$ to 3N generalized coordinates $\underset{\sim}{q}$ which can be divided into 3N-ℓ soft variables $\underset{\sim}{q}^{\alpha}$ and ℓ hard variables $\underset{\sim}{q}^{\beta}$. Forces in the system are conservative, i.e., the potential energy is a function of $\underset{\sim}{q}$ only. Momenta conjugate to q_i are defined as

$$p_i = \frac{\partial T}{\partial q_i} \tag{3.1}$$

where T is the total kinetic energy of the system.

The equilibrium average of an observable A is given by

$$\langle A \rangle = \frac{\int A \exp\{-\beta H(\underset{\sim}{p},\underset{\sim}{q})\} d\underset{\sim}{p} d\underset{\sim}{q}}{\int \exp\{-\beta H(\underset{\sim}{p},\underset{\sim}{q})\} d\underset{\sim}{p} d\underset{\sim}{q}} \tag{3.2}$$

This is true for any choice of canonical variables $\underset{\sim}{p},\underset{\sim}{q}$ in full phase space since the volume element $d\underset{\sim}{p}d\underset{\sim}{q}$ is invariant for canonical transformations.

We consider two cases*) (nomenclature according to [13]):

a. Flexible model: hard variables $\underset{\sim}{q}^{\beta}$ are subject to harmonic potentials in the limit of high force constants (essentially a Born-Oppenheimer approximation is used)

$$V(\underset{\sim}{q}) = V_s(\underset{\sim}{q}^{\alpha},\underset{\sim}{q}^{\beta}=\underset{\sim}{b}) + V_h(\underset{\sim}{q}^{\beta}) \tag{3.3}$$

$$V_h(\underset{\sim}{q}^{\beta}) = \frac{1}{2}(\underset{\sim}{q}^{\beta}-\underset{\sim}{b})^T \underset{\approx}{F}(\underset{\sim}{q}^{\beta}-\underset{\sim}{b}) \tag{3.4}$$

where $\underset{\approx}{F}$ is a matrix of force constants.

We will integrate in (3.2) over the variables $\underset{\sim}{p}^{\alpha},\underset{\sim}{p}^{\beta}$, and $\underset{\sim}{q}^{\beta}$. This will reduce (3.2) to integrals over $d\underset{\sim}{q}^{\alpha}$ only.

b. Rigid model: hard variables and their conjugate momenta are removed from the hamiltonian. Subsequent integration over $d\underset{\sim}{p}^{\alpha}$ also reduces (3.2) to integration over $d\underset{\sim}{q}^{\alpha}$. Results will be compared for the two cases.

*) Notation: In matrix notation we use $\underset{\sim}{a}$ for a column matrix representing the vector a; $\underset{\sim}{a}^T$ is a row matrix. Thus $\underset{\sim}{a}^T\underset{\sim}{b}$ is the scalar dot

product of vectors a and b. Integration variable da is shorthand for Πda_i. $|\underset{\approx}{A}|$ is the determinant of matrix. For simplicity we avoid the use of general co- and contravariant tensor notation. In consulting the literature one should be aware of the conflicting use of G on H in refs [13] and [14].

In order to determine the conjugate momenta we must write $T=\frac{1}{2}\sum_{k=1}^{3N} m_k \dot{x}_k^2$ in terms of $\underset{\sim}{q}$:

$$T = \frac{1}{2}\dot{\underset{\sim}{q}}^T \underset{\approx}{G}\dot{\underset{\sim}{q}} \tag{3.5}$$

where

$$G_{ij} = \sum_{k=1}^{3N} m_k \frac{\partial x_k}{\partial q_i}\frac{\partial x_k}{\partial q_j} \tag{3.6}$$

The symmetric square matrix $\underset{\approx}{G}$ we will call the "mass-metric tensor", in distinction with the metric tensor

$$q_{ij} = \sum_{k=1}^{3N} \frac{\partial x_k}{\partial q_i}\frac{\partial x_k}{\partial q_j} \tag{3.7}$$

which determines the path element in q-space:

$$(ds)^2 = d\underset{\sim}{q}^T \underset{\approx}{g}\, d\underset{\sim}{q} \tag{3.8}$$

and which plays a role in the direct transformation of cartesian into q-space.

From (3.1) and (3.5) we derive

$$\underset{\sim}{p} = \underset{\sim}{G}\dot{\underset{\sim}{q}} \tag{3.9}$$

and hence (with (3.5)

$$T = \frac{1}{2}\underset{\sim}{p}^T \underset{\sim}{G}^{-1}\underset{\sim}{p} \tag{3.10}$$

Integration over $d\underset{\sim}{p}$ yields

$$\int d\underset{\sim}{p}\, \exp\left\{-\frac{\beta}{2}\underset{\sim}{p}^T \underset{\approx}{G}^{-1}\underset{\sim}{p}\right\} = \left(\frac{2\pi}{\beta}\right)^{3N/2} |\underset{\approx}{G}|^{\frac{1}{2}} \tag{3.11}$$

Integration over dq yields, with (3.4):

$$\int d\underset{\sim}{q}^\beta\, \exp\left\{-\frac{\beta}{2}(\underset{\sim}{q}^\beta-\underset{\sim}{b})^T \underset{\approx}{F}(\underset{\sim}{q}^\beta-\underset{\sim}{b})\right\} = \left(\frac{2\pi}{\beta}\right)^{\ell} |\underset{\approx}{F}|^{-\frac{1}{2}} \tag{3.12}$$

Carrying out these integrations in both integrals of (3.2), constant terms cancel, including $|\underset{\approx}{F}|$ which we assumed to be independent of $\underset{\sim}{q}^\alpha$, and we obtain for the flexible molecule

$$\langle A\rangle = \frac{\int A|\underset{\approx}{G}|^{\frac{1}{2}} \exp\{-\beta V(\underset{\sim}{q}^\alpha)\}d\underset{\sim}{q}^\alpha}{\int |\underset{\approx}{G}|^{\frac{1}{2}} \exp\{-\beta V(\underset{\sim}{q}^\alpha)\}d\underset{\sim}{q}^\alpha} \tag{3.13}$$

We remark that, for this case of full dimensionality, in (3.13) the determinant of the metric tensor $|g|$ (3.7) may be used instead of $|\underset{\approx}{G}|$, because $|\underset{\approx}{G}| = \sum_{k=1}^{3N} m_k \cdot |g|$ and the mass product cancels. In its form with $|g|$, (3.13) can be derived directly from (3.2) in cartesian space by first integrating over momenta. Subsequent transformation from $\underset{\sim}{x}$ to $\underset{\sim}{q}$ then yields (3.13) because $d\underset{\sim}{x} = |\underset{\approx}{g}|^{\frac{1}{2}}\, d\underset{\sim}{q}$.

For the <u>rigid model</u> the treatment is similar. The massmetric tensor is now limited to the soft variables:

$$G^{\alpha}_{ij} = \sum_{k=1}^{3N} m_k \frac{\partial x_k}{\partial q^{\alpha}_i} \frac{\partial x_k}{\partial q^{\alpha}_j} \tag{3.14}$$

After integration over $\underset{\sim}{p}^{\alpha}$ we obtain

$$\langle A^{\alpha}\rangle = \frac{\int A|\underset{\approx}{G}^{\alpha}|^{\frac{1}{2}} \exp\{-\beta V(\underset{\sim}{q}^{\alpha})\}d\underset{\sim}{q}^{\alpha}}{\int |\underset{\approx}{G}^{\alpha}|^{\frac{1}{2}} \exp\{-\beta V(\underset{\sim}{q}^{\alpha})\}d\underset{\sim}{q}^{\alpha}} \tag{3.15}$$

Thus, a constraint simulation does not yield the same physical equilibrium averages unless $|G^{\alpha}| = |G|$. One may correct the constraint case by multiplying distribution functions by a factor

$$(|\underset{\approx}{G}|/|\underset{\approx}{G}^{\alpha}|)^{\frac{1}{2}} \tag{3.16}$$

Straightforward evaluation of $|G|$ and $|G^{\alpha}|$ is generally prohibitively complicated, although $|G|$ is quite simple for polymer chains [13]. Fixman [14] has derived the simple formula

$$|\underset{\approx}{G}^{\alpha}|/|\underset{\approx}{G}| = |\underset{\approx}{H}| \tag{3.17}$$

where $|\underset{\approx}{H}|$ is the determinant of $\underset{\sim}{H}^{\beta}$:

$$H^{\beta}_{ij} = \sum_{k=1}^{3N} m_k \frac{\partial q^{\beta}_i}{\partial x_k} \cdot \frac{\partial q^{\beta}_j}{\partial x_k} \tag{3.18}$$

This matrix is much more easily evaluated.

The correction can be put in the form of an additional potential

$$V' = \tfrac{1}{2}kT\, \ell n|\underset{\approx}{H}| \tag{3.19}$$

and then be used for dynamics simulations as well. This potential is the free energy due to the entropy $-\frac{1}{2}k\, \ell n|\underset{\approx}{H}|$ associated with harmonic oscillators in the constrained degrees of freedom.

Metric tensor corrections turn out to be non-negligable in the case of angle constraints [4,17,18]. For butane the corrections are not large for the trans and gauche minima, but quite considerable for the barrier in the dihedral potential. Uncorrected dynamics with constrained bond lengths and angles will overestimate transition probabilities between dihedral conformations by 20 to 30% [17]. The distribution function of the dihedral angle in the case that the internal dihedral potential is set to zero, is far from homogeneous when no corrections are made [18].

However, when only bond lenths are constrained [18] the metric effect is negligable. Although in principle the bond angle distribution receives a weight factor, the influence on the actual distribution is small because of the small range to which the valence angles are restricted. The effect of the compensating potential is to increase bond angles by 0.1 to 0.2^{o}. The metric tensor can be safely ignored in this case.

We may conclude that metric tensor corrections should be made if bond angles are constrained, but they can be ignored if only bond lengths are constrained.

4. CONCLUSIONS

For macromolecular dynamics we may conclude that constraining bond lengths only has a negligable effect on the physical behaviour of the system and significantly improves the computational efficiency. Metric tensor corrections are not necessary in this case. On the other hand, constraining bond angles in addition to bond lengths does not further improve computational efficiency; it does affect the physics of the system and metric tensor corrections will be significant.

This conclusion leading to the recommendation of bond length constraints only, is applicable to the range of force constants typical for polymers and proteins. The criterion is possible overlap and coupling between hard and soft modes. For small molecules bond deformation modes will in general be closer to bond stretching modes and better separated from intermolecular motions than in the case of polymers where coupling to other intramolecular modes occurs. Thus in small molecules it may be quite advantageous to constrain both bond lengths <u>and</u> angles. Tests such as described in Section 2 should be made in individual cases to assess the feasibility and effects of the incorporation of constraints. The metric tensor problem does not occur in rigid bodies without internal degrees of freedom.

REFERENCES

1. J.P. Ryckaert, G. Ciccotti and H.J.C. Berendsen, J. Comput.Phys. 23 (1977) 327.
2. H. Goldstein, Classical Mechanics, Addison-Wesley, Reading, Mass., 1950.

3. J. Wittenburg, Dynamics of Systems of Rigid Bodies, Teubner, Stuttgart 1977.
4. M.R. Pear and J.H. Weiner, J.Chem.Phys. 71 (1979) 212.
5. M. Fixman, J.Chem.Phys. 69 (1978) 1527.
6. L. Verlet, Phys.Rev. 159 (1967) 98.
7. J.P. Ryckaert and A. Bellemans, Faraday Discussions Chem.Soc. 66 (1978) 95.
8. W.F. van Gunsteren, H.J.C. Berendsen and J.A.C. Rullmann, Faraday Discussions Chem. Soc. 66 (1978) 58.
9. W.F. van Gunsteren and H.J.C. Berendsen, Mol.Phys. 34 (1977) 1311.
10. W.F. van Gunsteren and M. Karplus, "The effect of constraints on the dynamics of macromolecules", to be published.
11. W.F. van Gunsteren and M. Karplus, J.Comput.Chem. 1 (1980) 266.
12. J.A. McCammon, B.R. Gelin and M. Karplus, Nature 267 (1977) 585.
13. N. Gō and H.A. Scheraga, Macromolecules 9 (1976) 535.
14. M. Fixman, Proc.Natl.Acad.Sci. USA 71 (1974) 3050.
15. E. Helfand, J.Chem.Phys. 71 (1979) 5000.
16. M. Gottlieb and R.B. Bird, J.Chem.Phys. 65 (1976) 2467.
17. D. Chandler and B.J. Berne, J.Chem.Phys. 71 (1979) 5386.
18. W.F. van Gunsteren, Mol.Phys. 40 (1980) 1015.

APPENDIX: The subroutine SHAKE

SHAKE corrects coordinates x' such that a given set of holonomic constraints is satisfied. Corrections are carried out by successive displacements in the direction of reference vectors x, weighted with the inverse mass of the particles. The procedure uses a constraint list LISTCT (N) in which all particle numbers of the second particle of each constraint pair are listed. The first particles of the pairs occur in numerical order; the index of LISTCT (N) where particle k occurs for the first time is indicated in a pointer list LISTPT (K). If a particle has no constraint neighbour, its corresponding entry in LISTCT is zero. As an example consider butane with 5 constraints: three bond lengths a and two angles constrained by second neighbour distances b. The pointer list now contains 4 entries (4 particles); the constraint list contains 6 entries (5 constraints and one zero).

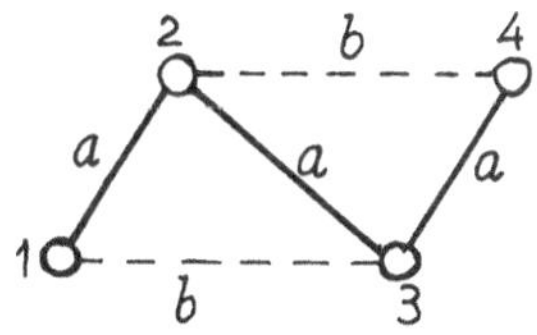

atom nr k = 1,4

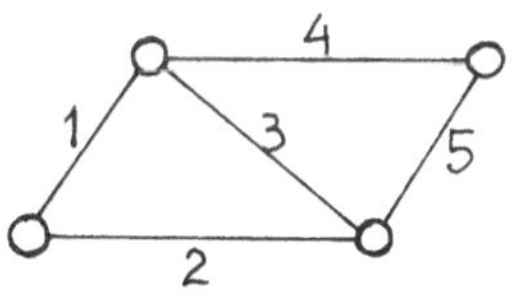

constraint nr n = 1,5

K	LISTPT (K)
1	1
2	3
3	5
4	6

N	LISTCT (N)	CONSTR (N)
1	2	a^2
2	3	b^2
3	3	a^2
4	4	b^2
5	4	a^2
6	0	-

The constraints are specified as the square of the constraint length in the parameter CONSTR (N).

If desired, formal parameters can be replaced by common variables.

We note that the appended FORTRAN subroutine is not optimal for use with vector- or array processors. In that case the tests on convergence, affecting the logical SKIP, should not be made after considering each constraint pair but after one iteration, involving all constraints, has been completed. The logical SKIP is not necessary in that case.

The present version is suitable for large polymers containing many different kinds of constraint.

When a large number identical particles is treated (such as a liquid of small rigid molecules), SHAKE should be called for each particle. It can be simplified considerably by implicit consideration of constraints without reference to constraint lists and pointer lists. For use with vector- or array processors the statements should then be applied to vectors involving all or many particles simultaneously.

```
CCCCCCCCCCCCCC-SUBR.SHAKE-H.J.C.BERENDSEN-UNIV.GRONINGEN-AUG 80-CCCCCCCC
C                                                                       C
C                                                                       C
C                                                                       C
C      subroutine SHAKE(NR,X,XPRIME,TOL,LISTPT,LISTCT,CONSTR,           C
C                     WINV,SKIP,NITER)                                  C
C                                                                       C
C                                                                       C
C            SHAKE will supply corrections to given coordinates        C
C     XPRIME such that for XPRIME a list of constraints will           C
C     be satisfied, each within a specified tolerance TOL. The         C
C     corrections are made along vectors derived from X.               C
C                                                                       C
C     Literature: Ryckaert, Ciccotti, Berendsen: J.Computat.Phys.      C
C     23 (1977) 327. The procedure is suitable for correction of a     C
C     molecular dynamics step using the Verlet Algorithm. For other    C
C     algorithms of the predictor-corrector type see: Van Gunsteren    C
C     and Berendsen Mol.Phys.34 (1977) 1311.                           C
C                                                                       C
C     The constraints are listed in the following order: First all     C
C     constraints connecting to particle 1, then all constraints       C
C     connecting to particle 2 (except the one between 1 and 2),       C
C     then all remaining constraints connecting to particle 3, etc.    C
C                                                                       C
C     NR is the number of atoms                                        C
C     TOL is the relative tolerance in each constraint distance.       C
C     LISTPT(k)(pointerlist) specifies the position in the             C
C        constraint list of the first constraint neighbour of          C
C        atom number k. The last constraint neighbour of atom k is     C
C        specified by LISTPT (k+1)-1. it is not necessary to define    C
C        LISTPT (nr+1).                                                C
C     LISTCT(n) (constraint list) specifies the atom numbers of the    C
C        constraint neighbours of k, starting at n=LISTPT(k) and       C
C        ending at LISTPT(k+1)-1. If atom k  has no constraint         C
C        neighbours, LISTCT(n)=0.                                      C
C     CONSTR(n) is the square of the length to which the distance      C
C        between atom k and the atom specified by LISTCT(n) is to      C
C        be constrained. It is not necessary to define CONSTR(n)       C
C        if LISTCT(n)=0                                                C
C     WINV(k) is the inverse of the mass of atom k                     C
C     SKIP is a dummy array of length 2*NR                             C
C     NITER (output) is the number of iterations                       C
C                                                                       C
C     If 100 or more iterations appear necessary, the subroutine       C
C     is returned with a message.                                      C
C     If XPRIME deviates too much from X, the subroutine is            C
C     returned with a message and NITER=0.                             C
C                                                                       C
C     Lines marked "DOUBLE" in columns 73-78 are specific for IBM      C
C     double precision. For single precision (CDC) remove the first    C
C     of these lines and change D into E in the others.                C
C                                                                       C
CCCCCCCCCCCCCCCCCCCCCCCCCCCCCCCCCCCCCCCCCCCCCCCCCCCCCCCCCCCCCCCCCCCCCCCCC
C
C
      subroutine SHAKE (NR,C,XPRIME,TOL,LISTPT,LISTCT,CONSTR,
     x                 WINV,SKIP, NITER)
      implicit REAL*8 (A-H,O-Z)                                         DOUBLE
      logical SKIP,READY
      dimension X(1),XPRIME(1),LISTPT(1),LISTCT(1),CONSTR(1),
     1          WINV(1),SKIP(1)
      NITER=0
      TOL2=2.D0*TOL                                                     DOUBLE
      do 10 K=1,NR
      SKIP(K)=.true.
      SKIP(NR+K)=.false.
   10 continue
      READY=.False.
   50 continue
      if(NITER.lt.100)goto 70
      print 61, TOL
   61 format (51H COORDINATE RESETTING (SHAKE) WAS NOT ACCOMPLISHED,
     1        33HWITHIN 100 ITERATIONS. TOLERANCE=,F10.6,/)
```

(continued)

Subroutine SHAKE (continued)

```
      return
   70 continue
      if (READY) return
      READY=.true.
      do 200 K=1,NR
      N1=LISTPT(k)
      if (LISTCT (N1).eq.0) goto 200
      KZ=3*K
      KY=KZ-1
      KX=KZ-2
      N2=LISTPT(K+1)-1
      do 100 N=N1,N2
      L=LISTCT(N)
      if (SKIP(NR+K).and.SKIP(NR+L)) goto 100
      LZ=3*L
      LY=LZ-1
      LX=LZ-2
      XDIF=XPRIME(KX)-XPRIME(LX)
      YDIF=XPRIME(KY)-XPRIME(LY)
      ZDIF=XPRIME(KZ)-XPRIME(LZ)
      DIFF=CONSTR(N)-XDIF*XDIF-YDIF*YDIF-ZDIF*ZDIF
      ABSDIF=DABS(DIFF)                                        DOUBLE
      TOLER=CONSTR(N)*TOL2
      if (ABSDIF.lt.TOLER) goto 100
      XKL=X(KX)-X(LX)
      YKL=X(KY)-X(LY)
      ZKL=X(KZ)-X(LZ)
      RRPR=XKL*XDIF+YKL*YDIF+ZKL*ZDIF
      if (RRPR.gt.1.D-6) goto 90                                DOUBLE
      print 75,K,N,NITER
   75 format(22H K , N  AND NITER ARE :,3I4)
      NITER=0
      print 81
   81 format(52H COORDINATE RESETTING (SHAKE) CANNOT BE ACCOMPLISHED,
     1       24H, DEVIATION IS TOO LARGE)
      return
   90 continue
      ACOR=.5D0*DIFF/RRPR*(WINV(K)+WINV(L)))                   DOUBLE
      XCRT=ACOR*XKL
      YCRT=ACOR*YKL
      ZCRT=ACOR*ZKL
      XPRIME(KX)=XPRIME(KX)+XCRT*WINV(K)
      XPRIME(KY)=XPRIME(KY)+YCRT*WINV(K)
      XPRIME(KZ)=XPRIME(KZ)+ZCRT*WINV(K)
      XPRIME(LX)=XPRIME(LX)-XCRT*WINV(L)
      XPRIME(LY)=XPRIME(LY)-YCRT*WINV(L)
      XPRIME(LZ)=XPRIME(LZ)-ZCRT*WINV(L)
      SKIP(K)=.false.
      SKIP(L)=.false.
      READY=.false.
  100 continue
  200 continue
      NITER=NITER+1
      do 250 K=1,NR
      SKIP(NR+K)=SKIP(K)
      SKIP(K)=.true.
  250 continue
      goto 50
      end
```

STOCHASTIC DYNAMICS OF POLYMERS

W.F. van Gunsteren and H.J.C. Berendsen

Laboratory of Physical Chemistry
University of Groningen
Nijenborgh 16
9747 AG Groningen
The Netherlands

ABSTRACT

When simulating a molecular system by the method of stochastic dynamics (SD), less interesting degrees of freedom are ignored and their influence on the other degrees of freedom is approximated by a combination of mean force interactions, stochastic forces and frictional forces. When time and space correlations in the latter are neglected, SD reduces to its simplest form, Brownian dynamics (BD). A BD algorithm is derived and applied to liquid n-butane and n-decane, where one molecule is considered explicitly and the surrounding liquid is modelled stochastically. From a comparison of the results to those of molecular dynamics (MD) simulations of these liquids it is concluded that the BD model yields a good approximation of the dynamics of n-alkanes in the liquid state.

1. INTRODUCTION

The computer simulation technique of molecular dynamics (MD) is a powerful tool for the description of molecular many-particle systems such as liquids and solutions. However, available computer power limits its application typically to systems containing several hundred to a few thousand atoms and to periods of time of the order of tens of picoseconds. In many instances one is only interested in certain aspects of the dynamics of the system (such as properties of solute molecules in a solution, dynamics of a specific particle in a lattice, intramolecular properties of a macromolecule in a bulk polymer, etc.). It would be highly advantageous if for such cases a method were available that correctly describes the dynamics of the relevant degrees of freedom while the less relevant degrees of freedom are not simulated in detail. The need for such methods is particularly pressing when the interest lies in long time scale events, for which "classical" MD would be completely prohibitive.

A case where the advantage of reduced methods is obvious, is the simulation of a macromolecule in a solvent, such as a protein in aqueous solution. In order to avoid non-realistic surface effects periodic boundary conditions are usually applied to liquids and solutions: the system is considered to be in a - generally rectangular - box, surrounded by identical copies of itself. When considering solutions of macromolecules, this procedure leads to the simulation of at least as many solvent molecules as there are atoms in the solute. For example, the protein BPTI (bovine pancreatic trypsin inhibitor) consisting of 454 heavy atoms needs about 2500 molecules of solvent if it is simulated in a rectangular periodic box with a minimum of two solvent layers between protein and walls of the box. In this case one is interested in the dynamics of the solute only and one would like to apply a reduced method retaining only a small fraction of the total number of degrees of freedom.

This is a problem that is generally encountered in any many-particle system: how to pick out the degrees of freedom that are essential for the phenomenon one wishes to describe, and how to model the interactions along these degrees of freedom such that they include the contribution of the (uninteresting) degrees of freedom that are subsequently neglected. In our case of the description of the dynamics of a macromolecule in solution the essential degrees of freedom are those of the solute. So, one may try to reduce or even to eliminate the solvent degrees of freedom while retaining a realistic simulation of the solute dynamics. In other cases the essential degrees of freedom are not so easily recognized; e.g. what is the reduced configurational space in which a chemical reaction in a catalytic process or the motion of a lattice interstitial in a crystal is described?

Here, we shall address the question of how the MD simulation of macromolecular solutions can be simplified by reducing or eliminating the solvent degrees of freedom and how the equations of motion of the remaining (solute) degrees of freedom should be modified in order to incorporate at least approximately the effect of the eliminated degrees of freedom of the solvent.

One approach is to simulate the solute in a box, the shape of which approximates the shape of the solute, but which also has the property that it covers together with the periodic copies of itself the whole 3-dimensional space. For example, a spherical solute could be simulated in a truncated octahedron, instead of in a cubic box, as has been suggested by Adams [1]. In this way the number of solvent molecules that are to be included in the simulation is proportional to R^2 instead of R^3 for the cubic box, R being the radius of the solute. A second approach is to neglect all solvent degrees of freedom; that is, the solute is simulated in vacuo [2]. In this zero order approximation of solvent effects the motion of atoms at the surface of the solute will definitely be simulated incorrectly. Therefore, one has to look for ways to incorporate the influence of the solvent molecules on the solute in the solute equations of motion without treating the solvent molecules explicitly. This can be done as follows.

1. A first order approximation of solvent effects is to use a potential of mean force, viz. a potential that includes in some way the average interaction with the solvent. It can be derived from a full-size MD simulation of solute plus solvent molecules in a box with periodic boundary conditions, or from analytical theories of the liquid state [3], or it may be conjectured in an empirical way from experimental solubility data [4].

2. A second order approximation of solvent effects is to apply stochastic dynamics (SD): the deviation of the real force from the mean force exerted by the solvent is represented by a randomly fluctuating force having stochastic properties that are determined by the solvent that is considered. These properties are generally represented in a simplified form. Parameters of the simplified representation are either derived from MD runs including all degrees of freedom or from experimental data, such as viscosity, diffusion coefficients or other transport properties.

In the following we review various levels of approximations used in SD. The remainder of this paper is written with a particular application in mind: the simulation of a polymer in a (stochastic) solvent; the methods of SD, however, can be easily adapted to a much wider variety of simulations.

Consider a system of N particles embedded in a medium (the "solvent") that weakly interacts with the system. The most simple form of SD is represented by the ordinary Langevin equation

$$m_i \dot{v}_i(t) = -m_i \gamma_i v_i(t) + F_i(\{x(t)\}) + R_i(t) \qquad (1.1)$$

together with the assumptions quoted below [5,6]. Here, i labels particles and components (i=1,2,...,3N). So, v_i is the x-,y-, or z-component of the velocity of a particle having mass m_i and friction coefficient γ_i. The systematic force F_i is to be derived from the potential (of mean force), which may depend on the coordinates of all particles of the system, denoted by $\{x(t)\}$. The random force is denoted by R_i. It is assumed to be stationary, Markovian and Gaussian with zero mean and to have no correlation with prior velocities:

$$\langle R_i(0)R_j(t)\rangle = 2m_i kT_o\gamma_i\delta_{ij}\delta(t) \qquad (1.2)$$

$$W_R(R_i) = (2\pi\langle R_i^2\rangle)^{-\frac{1}{2}} \exp(-R_i^2/2\langle R_i^2\rangle) \qquad (1.3)$$

$$\langle R_i\rangle = 0 \qquad (1.4)$$

$$\langle v_i(0)R_j(t)\rangle = 0 \qquad (1.5)$$

Here $\langle ...\rangle$ denotes averaging over an equilibrium ensemble and $W_R(R_i)$ is the (Gaussian) probability distribution of the random force. Since we are considering a stationary process, the ensemble average can also be obtained by time averaging. The coefficient of the delta-function in (1.2), or equivalently, the width of the Gaussian distribution, is determined by the second fluctuation-dissipation theorem [6], which takes care of balancing the increase in energy due to the randomly fluctuating force and the decrease in energy resulting from the friction force, keeping the molecule at the reference temperature T_o. Note that it is implicitly assumed in (1.1-5) that there is no feedback of the motion of the particles on the stochastic properties of the force exerted by the solvent.

This simplest form of SD is called Brownian dynamics (BD). It has been applied to a variety of systems, see e.g. [7] and refs. quoted there. A subclass of BD models is obtained by neglecting the inertial term (left hand side) in (1.1), which is physically justified when the viscosity is high, viz. when the physical phenomenon that is considered is essentially diffusive. Examples of this socalled diffusive Brownian dynamics (DBS) can be found in the refs. of [7].

The BD treatment can be generalized along two lines.

1. Time correlations are taken into account in generalized Langevin dynamics (GLD) [8], based on the generalized Langevin equation [6]

$$m_i\dot{v}_i(t) = -m_i\int_o^t \gamma_i(t')v_i(t-t')dt' + F_i(\{x(t)\}) + R_i(t) \qquad (1.6)$$

where the friction coefficient γ_i has become time-dependent. $R_i(t)$ in this case is not Markovian; its time correlation is related to $\gamma_i(t)$. Algorithms for integrating (1.6) are known [9]. Computer simulation of GLD has mainly been applied to model gas-solid collisions [8,10].

2. Spatial correlations can be taken into account by making the friction coefficient γ_i spacedependent, for example by writing

$$m_i\dot{v}_i(t) = m_i \sum_{j=1}^{3N} \gamma_{ij} v_j(t) + F_i(\{x(t)\}) + \sum_{j=1}^{3N} \alpha_{ij} R_j(t) \qquad (1.7)$$

This generalization is still to be explored. The spatial correlation between friction coefficients can be approximated by the assumption that the solvent behaves as a hydrodynamic continuum. Only a few studies of BD with hydrodynamic interactions for molecular systems are known [11,12].

Obviously, space and time correlations can both be taken into account. Thus far this has not been attempted.

It is not yet clear whether either or both of these two generalizations are essential when simulating a molecule in solution by SD; that is, whether time or space correlations play a significant role in the dynamics of the solute. The BD model completely ignores time and space correlations in the stochastic properties of the solvent. A way to check the validity of this simple model for the description of a molecule in solution is to compare the results of a BD simulation of the solute with those of a full-size MD simulation of the same solute plus solvent molecules. Here, this will be done for two systems, for which MD results were available at the time this study was started, viz. liquid n-butane and n-decane [13,14].

An algorithm for integrating the Langevin equation (1.1) is derived in section 2. The BD and MD results for n-butane and n-decane are compared in section 3. Section 4 contains some conclusions. A more detailed account of the investigations presented here, can be found in refs. [7,15].

2. ALGORITHMS FOR BROWNIAN DYNAMICS

Solving the linear, inhomogeneous first order differential equation (1.1), one finds

$$v(t) = v(0)e^{-\gamma t} + \int_0^t e^{-\gamma(t-t')}\{F(t')+R(t')\}m^{-1}dt' \qquad (2.1)$$

where the indices denoting particles and Cartesian components have been omitted. Various algorithms for BD simulations have been derived from (2.1). Turq et.al. [5] take the stochastic force R constant during the BD time step Δt. In this case the correlation time of the stochastic force is of the order Δt. In order to generate the stationary Markovian Gaussian process (1.1-5) the algorithm is limited by the condition

$$\Delta t << \gamma^{-1} \tag{2.2}$$

that is, that the correlation time of the stochastic force is much smaller than the velocity relaxation time γ^{-1}. Most algorithms used in the literature also require (2.2) to be valid. However, the integration time step Δt need not to be restricted by γ, since the influence of the stochastic force can be integrated over Δt [16]. So, Δt is only limited by the rate of change of the systematic force. When evaluating the integral in (2.1) we approximate $F(t')$ for $0 < t' < t$ by

$$F(t') = F(0) + \dot{F}(0)t' \tag{2.3}$$

where

$$\dot{F}(0) = \left.\frac{dF(t')}{dt'}\right]_{t'=0} \quad . \tag{2.4}$$

Using (2.3) in (2.1) we will derive a BD algorithm of the order $(\Delta t)^3$. We note that the efficient Verlet-algorithm [17], which is widely used in MD simulations, is also of order $(\Delta t)^3$. Therefore, we do not include higher order terms in (2.3). Going beyond third order would only pay when dealing with highly harmonic potentials [18]. A second order algorithm can be obtained by neglecting the second term in the right hand side of (2.3). In that case the algorithm does not "see" the curvature of the potential ($\dot{F}(t')=0$). This generates systematic errors in the integration of the equations of motion, whenever the potential is predominantly concave ($\dot{F}(t')>0$) or predominantly convex ($\dot{F}(t')<0$), as is often the case in molecular systems.

Evaluating the integral over the systematic force in (2.1) using (2.3), we find

$$\begin{aligned} v(t) &= v_o e^{-\gamma t} \\ &+ (m\gamma)^{-1}\{F(0)[1-e^{-\gamma t}] + \dot{F}(0)[t-\gamma^{-1}(1-e^{-\gamma t})]\} \\ &+ m^{-1}\int_o^t e^{-\gamma(t-t')}R(t')dt' \end{aligned} \tag{2.5}$$

and for the positions we get

$$\begin{aligned} x(t) &= x(0) + \int_o^t v(t')dt' \\ &= x(0) + v(0)\gamma^{-1}[1-e^{-\gamma t}] + (m\gamma)^{-1}\{F(0)[t-\gamma^{-1}(1-e^{-\gamma t})] \\ &+ \dot{F}(0)[\tfrac{1}{2}t^2-\gamma^{-1}[t-\gamma^{-1}(1-e^{-\gamma t})]]\} \\ &+ (m\gamma)^{-1}\int_o^t (1-e^{-\gamma(t-t')})R(t')dt' \quad . \end{aligned} \tag{2.6}$$

We define the random variables

$$X(t) \equiv (m\gamma)^{-1}\int_0^t (1-e^{-\gamma(t-t')})R(t')dt' \tag{2.7}$$

and

$$V(t) \equiv m^{-1}\int_0^t e^{-\gamma(t-t')}R(t')dt' \quad . \tag{2.8}$$

Since the stochastic force R(t') is a stationary, Markovian, Gaussian (with mean zero) stochastic process, X(t) and V(t) will have these same properties [19]. Since they are correlated by the fact that both depend on R(t'), they obey a bivariate Gaussian distribution [19].

$$w_{XV}(X,V) = [4\pi^2\sigma_1^2\sigma_2^2(1-r^2)]^{-\frac{1}{2}} \cdot \exp\{-[X^2/\sigma_1^2 - 2rXV/\sigma_1\sigma_2 + V^2/\sigma_2^2]/[2(1-r^2)]\} \quad . \tag{2.9}$$

The parameters σ_1, σ_2 and r of the distribution can be determined by evaluating the quantities $\langle X^2(t)\rangle$, $\langle V^2(t)\rangle$ and $\langle X(t)V(t)\rangle$ using (2.7-8) and the property (1.2)

$$\langle R(t')R(t'')\rangle = 2mkT\gamma\delta(t'-t'') \quad . \tag{2.10}$$

We find

$$\sigma_1^2 = \langle X^2(t)\rangle = kT(m\gamma)^{-1}C(t) \quad , \tag{2.11}$$

$$\sigma_2^2 = \langle V^2(t)\rangle = kTm^{-1}\{1-e^{-2\gamma t}\} \quad , \tag{2.12}$$

$$r\sigma_1\sigma_2 = \langle X(t)V(t)\rangle = kT(m\gamma)^{-1}\{1-e^{-\gamma t}\}^2 \quad , \tag{2.13}$$

$$\sigma_2^2(1-r^2) = \sigma_2^2 - \left(\frac{r\sigma_1\sigma_2}{\sigma_1}\right)^2 = 2kTm^{-1}B(t)/C(t) \quad , \tag{2.14}$$

and

$$r\sigma_2/\sigma_1 = r\sigma_1\sigma_2/\sigma_1^2 = \gamma(1-e^{-\gamma t})^2/C(t) \tag{2.15}$$

where

$$B(t) \equiv \gamma t(1-e^{-2\gamma t}) - 2(1-e^{-\gamma t})^2 \tag{2.16}$$

and

$$C(t) \equiv 2\gamma t - 3 + 4e^{-\gamma t} - e^{-2\gamma t} \tag{2.17}$$

Sampling in a bivariate distribution like (2.9) is generally performed by first sampling one variable from its distribution, and subsequently sampling the other variable from its conditional distribution given the obtained value of the first variable [19]. The distribution for X(t), irrespective the value of V(t), is

$$w_X(X(t)) = (2\pi\sigma_1^2)^{-\frac{1}{2}} \exp\{-X^2(t)/2\sigma_1^2\} \tag{2.18}$$

and the conditional distribution for V(t), given a specific value X=X(t) reads [19]

$$w_V(V(t)\,|X=X(t)) =$$
$$\{2\pi\sigma_2^2(1-r^2)\}^{-\frac{1}{2}} \exp\{-[V(t)-r\sigma_2\sigma_1^{-1}X(t)]^2/[2\sigma_2^2(1-r^2)]\} \tag{2.19}$$

From (2.5-8) and (2.18-19) we can now derive an algorithm in which two independent Gaussian variables occur. To 2nd order in Δt (omitting $\dot{F}$) it is equivalent to an algorithm derived by Ermak and Buckholtz [16] from the Fokker-Planck equation.

$$\begin{aligned} x_{n+1} &= x_n + v_n\gamma^{-1}(1-e^{-\gamma\Delta t}) + \\ &+ (m\gamma)^{-1}\{F_n[\Delta t-\gamma^{-1}(1-e^{-\gamma\Delta t})] + \\ &\qquad + \dot{F}_n[\tfrac{1}{2}(\Delta t)^2-\gamma^{-1}[\Delta t-\gamma^{-1}(1-e^{-\gamma\Delta t})]]\} + \\ &+ X_n(\Delta t) , \end{aligned} \tag{2.20}$$

$$\begin{aligned} v_{n+1} &= v_n e^{-\gamma\Delta t} + \\ &+ (m\gamma)^{-1}\{F_n(1-e^{-\gamma\Delta t}) + \dot{F}_n[\Delta t-\gamma^{-1}(1-e^{-\gamma\Delta t})]\} + \\ &+ r\sigma_2\sigma_1^{-1}\{x_{n+1}-x_n-v_n\gamma^{-1}(1-e^{-\gamma\Delta t}) + \\ &\qquad - (m\gamma)^{-1}\{F_n[\Delta t-\gamma^{-1}(1-e^{-\gamma\Delta t})] + \\ &\qquad\quad + \dot{F}_n[\tfrac{1}{2}(\Delta t)^2-\gamma^{-1}[\Delta t-\gamma^{-1}(1-e^{-\gamma\Delta t})]]\}\} + \\ &+ V_n(\Delta t) . \end{aligned} \tag{2.21}$$

Here, x_n is a short hand notation for $x(t_n)$. The random variables $X_n(\Delta t)$ and $V_n(\Delta t)$ are being sampled from two independent Gaussian distributions, characterized by

$$\langle X_n(\Delta t)\rangle = \langle V_n(\Delta t)\rangle = \langle X_n(\Delta t)V_n(\Delta t)\rangle = 0 \quad , \tag{2.22}$$

$$\langle X_n^2(\Delta t)\rangle = \sigma_1^2 = kT_o(m\gamma^2)^{-1}\, C(\Delta t) \tag{2.23}$$

and

$$\langle V_n^2(\Delta t)\rangle = \sigma_2^2(1-r^2) = 2kT_o m^{-1} B(\Delta t)/C(\Delta t) \quad . \tag{2.24}$$

The reference temperature is denoted by T_o.

Finally, the computational scheme for BD simulations using this algorithm looks as follows.

1. Assume that x_n, v_n and F_{n-1} are known.

2. Evaluate F_n from the potential $V(\{x_n\})$.

3. Compute the derivative of the systematic force from

$$\dot{F}_n = (F_n - F_{n-1})/\Delta t \tag{2.25}$$

4. Sample $X_n(\Delta t)$ using (2.23) and calculate the positions x_{n+1} from (2.20).

5. Sample $V_n(\Delta t)$ using (2.24) and calculate the velocities v_{n+1} from (2.21).

In the first step of a BD run we take $\dot{F}_o=0$ instead of using (2.25), since F_{-1} is not known.

If we put $\dot{F}_n=0$ in (2.20-21) the third order algorithm reduces to the second order one proposed in [16]. If we let γ go to zero in (2.20-21) we find the MD algorithm

$$x_{n+1} = x_n + v_n\Delta t + m^{-1}\{F_n \tfrac{1}{2}(\Delta t)^2 + \dot{F}_n \tfrac{1}{6}(\Delta t)^3\} , \tag{2.26}$$

$$v_{n+1} = v_n + m^{-1}\{F_n\Delta t + \dot{F}_n \tfrac{1}{2}(\Delta t)^2\} \tag{2.27}$$

which is equivalent to Verlet's algorithm [17]. For large values of γ, in the diffusive regime, when the friction is so strong that the velocities relax within t, eqs. (2.20-24) reduce to

$$x_{n+1} = x_n + (m\gamma)^{-1}\{F_n\Delta t + \dot{F}_n \tfrac{1}{2}(\Delta t)^2\} + X_n(\Delta t) , \tag{2.28}$$

$$\langle X_n(\Delta t)\rangle = 0 , \tag{2.29}$$

$$\langle X_n^2(\Delta t)\rangle = 2kT_o(m\gamma)^{-1} \Delta t . \tag{2.30}$$

When constraints such as bond-length or bond-angle constraints, are introduced in the molecule, the computational scheme has to be changed. This has been discussed in [15].

3. TEST OF THE BD DESCRIPTION OF LIQUID N-ALKANES

The assumptions underlying the BD model can be evaluated by comparing our BD results to those of the MD simulations of liquid n-butane and n-decane by Ryckaert and Bellemans [13,14].

3.1 Model for the n-alkanes and computational parameters

The model for the n-alkanes that we use is identical to that of refs. [13,14]. This means that we do not apply a potential of mean force, which enables us to evaluate the role of the mean force when comparing BD and MD results.

The concept of united atoms is used, so hydrogen atoms are not explicitly considered, but incorporated into the C-atoms to which they are bound. The bond lengths are constrained to the value of 0.153 nm and the bond angles are kept tetrahedral. These constraints are conserved by applying the procedure SHAKE [20] with a relative accuracy of 10^{-5}. No metric tensor potential is included, since it has also been omitted in the MD simulations of refs. [13,14] to which our BD results will be compared. The masses of all united atoms have been taken equal to m = 14 u. The torsional potential for the dihedral angles φ is given by

$$\begin{aligned} V(\varphi) = {} & 9.2789 + 12.1557 \cos\varphi - 13.1201 \cos^2\varphi + \\ & - 3.0597 \cos^3\varphi + 26.2403 \cos^4\varphi + \\ & - 31.4950 \cos^5\varphi \ \text{kJ mol}^{-1} \end{aligned} \tag{3.1}$$

United atoms that are at least four bonds apart interact through a Lennard-Jones potential with the parameters

$$\varepsilon = 0.5986 \ \text{kJ.mol}^{-1} , \qquad = 0.3923 \ \text{nm} . \tag{3.2}$$

We use the units kJ mole^{-1} (energy), u (mass: atomic lass unit), nm (length) and K (temperature), which implies that the unit of time is ps. The reference temperature T_o for the BD run is taken equal to the average temperature of the MD run to which it will be compared; that is, T_o = 291.5 K for n-butane and T_o = 481.0 K for n-decane [13,14]. The atomic friction coefficient γ is chosen such that the BD molecular diffusion constant is equal to the MD one; that is, $\gamma = 6.658$ ps^{-1} for n-butane and $\gamma = 3.759$ ps^{-1} for n-decane [7]. The BD runs for n-butane covers 1 ns and that for n-decane 600 ps.

3.2 BD versus MD for n-butane

The BD φ-distribution is almost identical to the gas-phase distribution, whereas the MD distribution shows an enhancement of the number of gauche conformations (table 1). This can be understood from packing or steric effects playing a much more significant role in the condensed phase than in the gas phase [21]. Clearly the

BD model cannot produce this enhancement. However, these packing effects could be taken into account in an approximate way by applying a potential of mean force [22].

The BD model implies by definition an exponential center of mass velocity autocorrelation function. It differs not too much from the MD result, suggesting that the neglect of time correlations in the friction yields a reasonable first order approximation of the dynamics on a molecular level.

Table I

Percentages of trans and gauche conformations for n-butane and n-decane

	BD		MD		gas phase	
	trans	gauche	trans	gauche	trans	gauche
n-butane φ	66	34	54	46	66	34
n-decane φ_1,φ_7	57	43	54	46	-	-
φ_2,φ_6	64	36	64	36	-	-
φ_3,φ_5	62	38	66	34	-	-
φ_4	54	46	55	45	-	-

3.3 BD versus MD for n-decane

For decane the fraction of trans versus gauche conformations for the different angles obtained by BD agrees well with the MD results, in contrast to the results for butane (table 1). This indicates that for longer chains the average conformation is more determined by the intramolecular interactions than by solvent packing effects.

The BD and MD center of mass velocity autocorrelation functions are rather similar (almost exponential) for decane, indicating that the time correlations in the friction play a minor role in the dynamics of the center of mass.

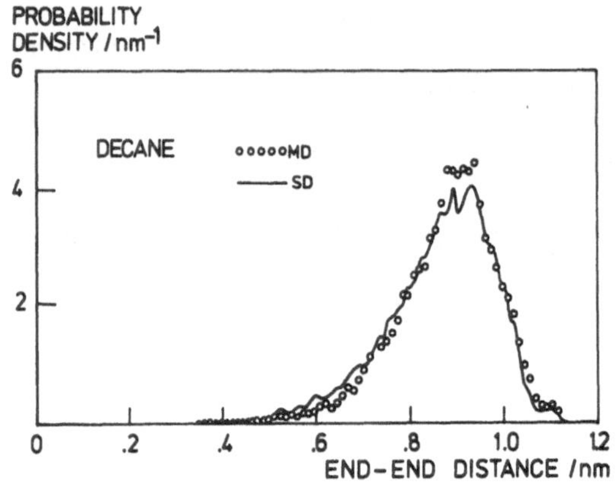

Fig. 1. Probability density of the end-to-end distance R.

The conformation of the decane molecule can globally characterized by its end-to-end distance R. The BD and MD distributions of R are displayed in fig. 1 and the various averages are given in table 2. BD and MD give almost identical results. The BD value for $\langle R \rangle$ is slightly smaller than the MD value, because in the former case there is no mean force exerted by the solvent molecules, which pulls the molecule slightly apart.

Table 2

Averages of the end-to-end distance (R) distribution

	BD	MD
$\langle R \rangle$	.864	.881
$\langle [R-\langle R \rangle]^2 \rangle^{\frac{1}{2}}$	.116	.109
$\langle R^2 \rangle$	.760	.787
$\langle [R^2-\langle R^2 \rangle]^2 \rangle^{\frac{1}{2}}$	.193	.181

The autocorrelation function of R^2 is given in fig. 2. The BD and MD functions exhibit the same period, but the latter are slightly more damped. Also the autocorrelation functions $\langle \cos[\varphi(t)-\varphi(0)] \rangle$ for the different angles are similar but show less damping in the BD case than in the MD case (fig. 3).

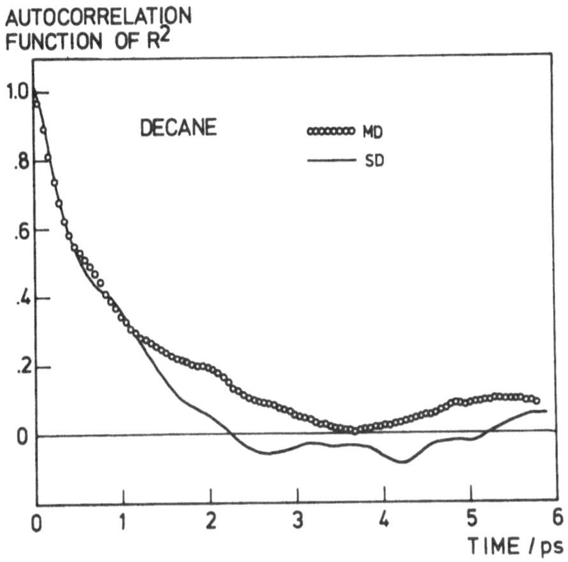

Fig. 2. Autocorrelation function of the square of the end-to-end distance R.

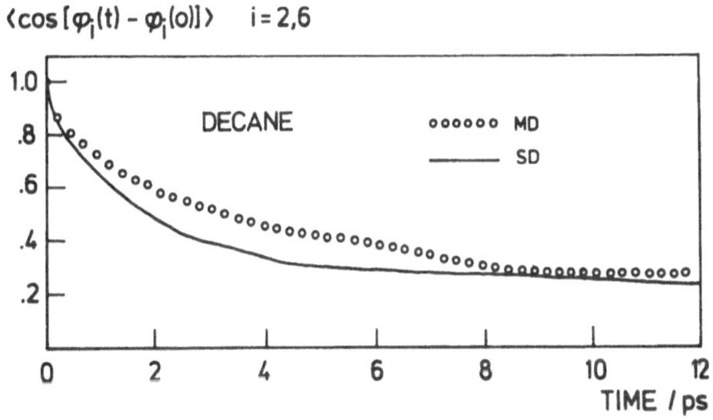

Fig. 3. Dihedral autocorrelation function.

So, autocorrelation of internal motions (R,cosφ) is not sufficiently damped. For this a larger γ seems necessary than is compatible with the diffusion constant of the center of mass. It is likely that this discrepancy disappears when hydrodynamic interaction is taken into account. Such interaction diminishes the rotational and translational friction of the molecule by introducing correlation in the frictional coefficient. Thus, when hydrodynamic interactions are included, γ will take on a larger value to account for the friction of internal motions.

Summarizing we conclude that the BD model yields a good first order approximation of the dynamics of a chain molecule in the liquid state. For butane, solvent packing effects play an important role in the condensed phase. However, for decane, the equilibrium conformation and dynamics are mainly determined by intramolecular interactions.

4. CONCLUSIONS

The BD model appears to yield a good approximation of the dynamics of n-alkanes in the liquid state. The stochastification of the solvent degrees of freedom leaves the characteristic features of the solute equilibrium properties and dynamics unchanged. When simulating a molecular solution, the majority of the degrees of freedom are those of the solvent. So, the computational effort of simulating a (macromolecular) solution may be reduced by at least an order of magnitude by applying the BD model instead of performing a full MD simulation of solute plus solvent. For example, a MD simulation of the protein BPTI in a Lennard-Jones solvent took about 20 hours CPU time per ps on a VAX 11-780, whereas a BD simulation would take less than one hour CPU time per ps. In addition, stochastic treatment of solvent degrees of freedom may often reduce the complexity of the simulation considerably. An example is the BD of electrolyte solutions [5] where the behaviour of solutions of molar concentration can be quite well modelled by Brownian description of the solvent.

The observation that time and space correlations in the force field exerted by the solvent play a minor role in the dynamics of the solute, is also computationally advantageous. The introduction of spatial correlations, e.g. as a hydrodynamic interaction, in a SD simulation will increase the computational effort considerably, since this type of interaction is essential longranged, which means that many atom pairs have to be considered when calculating the hydrodynamic interaction. However, it remains to be investigated whether time or space correlations play also a minor role at other temperatures or viscosities, or in other systems. We are presently testing the BD model for BPTI in a solvent consisting of Lennard-

Jones atoms. The application of the BD model to protein dynamics is especially interesting since there is experimental evidence that the solvent viscosity influences the rate of reaction steps that take place inside the protein [23].

Finally, we note that the idea and technique of stochastification of physically less interesting degrees of freedom of a many-particle system in order to reduce the size of the system and to localize the essential degrees of freedom, is not only valuable in the field of molecular solutions, but will find applications in a wide variety of phenomena. Among these will no doubt be the study of activated processes such as chemical reactions. When a reactive system can be described in a low-dimensional space, and stochastic properties are known, the rate of the activated processes can be accurately predicted by available theories. We have found this to be true for the trans-gauche transition rates in butane and decane [7]. It is also likely that stochastic models will be applicable to dynamic properties in the solid state, such as superionic conduction, diffusion and defect dynamics.

REFERENCES

1. D.J. Adams, Chem. Phys. Letters 62 (1979) 329.
2. J.A. McCammon, B.R. Gelin and M. Karplus, Nature 267 (1977) 585.
3. L.R. Pratt and D. Chandler, J.Chem.Phys. 67 (1977) 3683.
4. F.M. Richards, Ann.Rev.Biophys.Bioeng. 6 (1977) 151.
5. P. Turq, F. Lantelme and H.L. Friedman, J.Chem.Phys. 66 (1977) 3039.
6. R. Kubo, Rep.Prog.Phys. 29 (1966) 255.
7. W.F. van Gunsteren, H.J.C. Berendsen and J.A.C. Rullmann, "Stochastic Dynamics for Molecules with Constraints, Brownian Dynamics of n-alkanes", submitted to Mol.Phys.
8. J.D. Doll and D.R. Dion, J.Chem.Phys. 65 (1976) 3762.
9. G. Cicotti and J.P. Ryckaert, Mol.Phys. 40 (1980) 141.
10. S.A. Adelman, J.Chem.Phys. 71 (1979) 4471.
11. D.L. Ermak and J.A. McCammon, J.Chem.Phys. 69 (1978) 1352.
12. J.S. McCaskill and R.G. Gilberg, Chem.Phys. 44 (1979) 389.
13. J.P. Ryckaert, "Simulation de n-alkanes liquides par la méthode de dynamique moléculaire", thesis, Université Libre de Bruxelles, Brussels, 1976.
14. J.P. Ryckaert and A. Bellemans, Faraday Disc.Chem.Soc. 66 (1978) 95.
15. W.F. van Gunsteren and H.J.C. Berendsen, "Algorithms for Brownian Dynamics", to be published.
16. D.L. Ermak and H. Buckholtz, "Numerical Integration of the Langevin Equation: Monte Carlo Simulation", preprint, 1978.
17. L. Verlet, Phys.Rev. 159 (1967) 98.

18. W.F. van Gunsteren and H.J.C. Berendsen, Mol.Phys. 34 (1977) 1311.
19. A. Papoulis, "Probability, Random Variables, and Stochastic Processes, McGraw-Hill, 1965.
20. J.P. Ryckaert, G. Ciccotti and H.J.C. Berendsen, J.Comput.Phys. 23 (1977) 327.
21. D. Chandler, Faraday Disc.Chem.Soc. 66 (1978) 184.
22. R.M. Levy, M. Karplus and J.A. McCammon, Chem.Phys.Letters 65 (1979) 4.
23. D. Beece et al., "Solvent Viscosity and Protein Dynamics", Biochemistry (1980) in press.

APPLICATION OF IRREVERSIBLE THERMODYNAMICS TO MASS TRANSPORT IN IONIC CONDUCTORS

Frederick H. Horne

Department of Chemistry
Michigan State University
East Lansing, MI 48824 U.S.A.

I. INTRODUCTION

Despite the almost universal acceptance of the non-equilibrium, or irreversible, thermodynamics (NET) of Onsager[1] and his followers,[2,3] most of the published research concerning ion transport relies on the Nernst-Planck equations.[4] The Nernst-Planck equations ignore cross-effects due to both equilibrium and non-equilibrium interactions. Cross-effects cannot be ignored in complete treatments of ion transport, as others[5-9] have also noted. For an ion of concentration c_i, charge z_i, and velocity v_i, the one dimensional Nernst-Planck equation is

$$c_i v_i = - D_i \left(\frac{\partial c_i}{\partial x} - c_i z_i \frac{F}{RT} E \right) \tag{1.1}$$

where D_i is a diffusion coefficient, F is the Faraday constant, R the gas constant, T the temperature, and E the electric field. The principal inaccuracy in Eq. (1.1) is the omission of terms which represent the effects upon $c_i v_i$ of interactions among ions of different kinds. The NET equation corresponding to (1.1) is, for n solute components,

$$c_i v_i = c_i v_0 - \sum_j D_{ij} \frac{\partial c_j}{\partial x} + \frac{\lambda t_i}{F Z_i} E , \quad i,j = 1, \ldots, n, \tag{1.2}$$

where v_0 is the velocity of the solvent (usually vanishingly small), λ is the specific conductance (often denoted σ by solid state physicists), and t_i is the Hittorf transference number for ion i in the particular solution. Eq. (1.2) reduces to Eq. (1.1) for very dilute solutions.

In addition to the ion flux equations above, a nonequilibrium ionic system also obeys the equations of conservation of mass, energy, and momentum. The exact local mass conservation equations for non-reacting systems are

$$\frac{\partial c_i}{\partial t} = -\frac{\partial}{\partial x} c_i v_i , \quad i = 0, \ldots, n. \tag{1.3}$$

The relationship of the electric field to concentration is assumed to be given by the Poisson equation,

$$\frac{\partial E}{\partial x} = \frac{4\pi F}{\varepsilon} \sum_i z_i c_i , \tag{1.4}$$

where ε is the permittivity of the medium. For computer simulation it is useful to use the displacement current equation,[10,11]

$$I = F \sum_i z_i c_i v_i + \frac{\varepsilon}{4\pi} \frac{\partial E}{\partial t} , \tag{1.5}$$

where I is the current and $(\varepsilon/4\pi)(\partial E/\partial t)$ is Maxwell's displacement current.[12]

Substitution of Eqs. (1.2) into Eqs. (1.3) and (1.5) and rearrangement yield the complete set of n + 1 transport equations for isothermal systems with $v_0 = 0$,

$$\begin{aligned} \frac{\partial E}{\partial t} &= \frac{4\pi}{\varepsilon} \left(I + F \sum_i \sum_j z_i D_{ij} \frac{\partial c_i}{\partial x} - \lambda E\right) , \\ \frac{\partial c_i}{\partial t} &= \frac{\partial}{\partial x} \left(\sum_j D_{ij} \frac{\partial c_j}{\partial x} - \frac{\lambda t_i}{F z_i} E\right). \end{aligned} \tag{1.6}$$

In the next section, Eqs. (1.2) are derived from the more fundamental Onsager equations. In section III, the concentration dependence of D_{ij}, λ, and t_i for binary electrolyte solutions is discussed with the help of the compilations of Miller.[13] This discussion leads to explicit statements about the range of validity of the Nernst-Planck equations and of other approximations. Some analytical and numerical approaches to solving the isothermal transport equations for various boundary conditions are discussed briefly in Section IV. The extension of the transport equations to nonisothermal systems is presented in Section V.

II. NET EQUATIONS

Hittorf diffusion fluxes, defined by

$$J_i^H = c_i v_i - c_i v_0 \,, \quad i = 1, \ldots, n, \tag{2.1}$$

are, according to the Onsager approach,[1,2] linear combinations of the gradients of the chemical potential μ_i in an isothermal system,

$$-J_i^H = \sum_{j=1}^{n} \ell_{ij} \frac{\partial \mu_j}{\partial x} \,. \tag{2.2}$$

The matrix of the Onsager coefficients ℓ_{ij} is symmetric,

$$\ell_{ij} = \ell_{ji} \,. \tag{2.3}$$

With composition expressed in molar units, c_i, the chemical (or electrochemical) potential for ion i is given by

$$\mu_i = \mu_i^{\ominus}(T,p) + RT\ell n\, c_i y_i + z_i F\phi \,, \tag{2.4}$$

where y_i is the molarity-based activity coefficient and ϕ is the electrostatic potential. The gradient of μ_i in terms of gradients of c_i and ϕ is then, for an isothermal, isobaric system,

$$\frac{\partial \mu_i}{\partial x} = \sum_{j=1}^{n} \mu_{ij} \frac{\partial c_j}{\partial x} + z_i F \frac{\partial \phi}{\partial x} \,, \tag{2.5}$$

with

$$\mu_{ij} = \left(\frac{\partial \mu_i}{\partial c_j}\right)_{T,p,c_{k\neq j}} = \frac{RT}{c_j}\left[\delta_{ij} + \left(\frac{\partial \ell n y_i}{\partial \ell n c_j}\right)_{T,p,c_k}\right] \,, \tag{2.6}$$

where δ_{ij} is the Kronecker delta. Moreover,

$$E = -\frac{\partial\phi}{\partial x} . \qquad (2.7)$$

Thus, Eqs. (2.2) become, for each ion,

$$-J_i^H = \sum_j D_{ij} \frac{\partial c_j}{\partial x} - \frac{\lambda t_i}{F z_i} E , \qquad (2.8)$$

with

$$D_{ij} = \sum_k \ell_{ik}\mu_{kj} , \quad t_i = \frac{F^2 z_i}{\lambda} \sum_k z_k \ell_{ik} ,$$
$$\lambda = F^2 \sum_i \sum_j z_i z_j \ell_{ij} , \quad \sum_i t_i = 1 . \qquad (2.9)$$

Note that Eq. (2.8) is identical to Eq. (1.2). In terms of partial conductances,[9] we have

$$\sigma_{ij} = z_i z_j \ell_{ij} F^2 , \quad \lambda = \sigma = \sum_i \sum_j \sigma_{ij} . \qquad (2.10)$$

In order for Eqs. (1.2) to reduce to the Nernst-Planck equation, Eqs. (1.1), it is necessary and sufficient that

(i) $v_0 = 0$,

(ii) $D_{ij} = D_i \delta_{ij}$, $i,j = 1, \ldots, n$,

and

(iii) $\frac{\lambda t_i}{F z_i} = c_i z_i D_i \frac{F}{RT}$, $i = 1, \ldots, n$.

Coupled with Eqs. (2.6) and (2.9), these restrictions imply that alternative conditions for validity of the Nernst-Planck equations are

(ii)' $\ell_{ij} = \ell_{ii}\delta_{ij}$, $i,j = 1, \ldots, n$,

(iii)' $\ell n y_i = 0$, $i=1, \ldots, n$.

Thus, the Nernst-Planck equations are valid only for

ideal solutions and for no cross-terms in the flux equations. Put another way, the Nernst-Planck equations are valid only when there is neither equilibrium interaction among molecules of different types nor interaction of flows among molecules of different types. In terms of the Nernst-Planck diffusion coefficients

$$D_i = RT \frac{\ell_{ii}}{c_i} , \tag{2.11}$$

the accurate flux equations are

$$c_i v_i = c_i v_0 - D_i \frac{\partial c_i}{\partial x} - (D_{ii} - D_i) \frac{\partial c_i}{\partial x} - \sum_{j \neq i} D_{ij} \frac{\partial c_j}{\partial x} + D_i c_i z_i \frac{F}{RT} E + FE \sum_{j \neq i} z_j \ell_{ij} , \tag{2.12}$$

with

$$D_{ij} = \frac{RT}{c_i} [\ell_{ij} + \sum_k \ell_{ij} (\frac{\partial \ell n y_k}{\partial \ell n c_i})], \quad i \neq j ,$$
$$D_{ii} = D_i + \frac{RT}{c_i} \sum_k \ell_{ik} (\frac{\partial \ell n y_k}{\partial \ell n c_i}) . \tag{2.13}$$

The so called "ambipolar" or "chemical" or "mutual" diffusion coefficient $\tilde{D}$ for a uni-univalent binary mixture is related to the D_{ij} by[7,9]

$$\tilde{D} = 2(D_{11}D_{22} - D_{12}D_{21})/(D_{11} + D_{22} - D_{21} - D_{12}). \tag{2.14}$$

For very dilute solutions (see section III), this reduces to

$$\tilde{D} = 2D_1 D_2/(D_1 + D_2) . \tag{2.15}$$

The thorough analysis by Huggins and Huggins[14] of the relationship between $\tilde{D}$ and the self-diffusion coefficients should be extended to include the coupled effects due to cross-coefficients and activity coefficients.

The mobility u_i (not to be confused with chemical potential μ_i) is defined by the Nernst-Einstein relation (See Haase, in Reference 2, p. 260)

$$u_i = |z_i| D_i F/RT = |z_i| \ell_{ii} F/c_i \, . \tag{2.16}$$

Howard and Lidiard[5] derived a generalized Nernst-Einstein relation which has not, unfortunately, been used extensively. In terms of the mobilities of Eq. (2.16), the correct expression for the conductance is

$$\begin{aligned} \lambda = \sigma &= F\sum_i |z_i| c_i u_i + F^2 \sum_i \sum_{j\neq i} z_i z_j \ell_{ij} \\ &= \sum_i |z_i| c_i u_i + \sum_i \sum_{j\neq i} \sigma_{ij} \, . \end{aligned} \tag{2.17}$$

Only the first summation, $\sum_i |z_i| c_i u_i$, appears in most works. The cross-coefficients ℓ_{ij}, $i\neq j$, must all vanish for $\sigma = \sum_i |z_i| c_i u_i$ to be correct.

TABLE I - Some Values[a] of Diffusivities[b] for Binary Aqueous Electrolyte Solutions.

c	D_{11}	D_{12}	D_{21}	D_{22}	D_1	D_2
			H_2O - $LiC\ell$			
0.01	0.95	0.01	-0.03	1.89	0.99	1.97
0.10	0.87	0.06	-0.01	1.74	0.94	1.87
1.00	0.86	0.30	0.42	1.80	0.73	1.56
			H_2O - $NaC\ell$			
0.01	1.23	--	-0.03	1.89	1.29	1.98
0.10	1.12	0.02	-0.03	1.72	1.23	1.88
1.00	1.04	0.20	0.19	1.64	1.07	1.68
			H_2O - $KC\ell$			
0.01	1.82	-0.03	-0.03	1.89	1.91	1.98
0.10	1.66	-0.03	-0.03	1.72	1.84	1.91
1.00	1.56	0.11	0.10	1.62	1.75	1.83

[a]Entries calculated from compilations of D.G. Miller, J. Phys. Chem. 70, 2639 (1966).

[b]See text for definitions of coefficients. Units of c are mol dm^{-3}; units of diffusivities are $cm^2 sec^{-1}$.

TABLE II - Some Values[a] of Conductances[b] and Transference Numbers[b] for Binary Aqueous Electrolyte Solutions.

c	$10^2\lambda = 10^2\sigma$	$10^2\sigma_{11}$	$10^2\sigma_{12}$	$10^2\sigma_{22}$	t_1	t_2
		H_2O - LiCℓ				
0.01	0.107	0.037	0.002	0.074	0.33	0.67
0.10	0.959	0.351	0.048	0.703	0.32	0.68
1.00	7.29	2.74	0.65	5.85	0.29	0.71
		H_2O - NaCℓ				
0.01	0.119	0.049	0.002	0.074	0.39	0.61
0.10	1.067	0.463	0.052	0.708	0.39	0.61
1.00	8.576	4.014	0.848	6.305	0.37	0.63
		H_2O - KCℓ				
0.01	0.141	0.072	0.002	0.074	0.49	0.51
0.10	1.290	0.692	0.060	0.718	0.49	0.51
1.00	11.187	6.589	1.130	6.857	0.49	0.51

[a] Entries calculated from compilations of D.G. Miller, J. Phys. Chem. 70, 2639 (1966).

[b] See text for definitions of coefficients. Units of c are mol·dm^3; units of conductances are ohm^{-1}cm^{-1}; the t_i are dimensionless. For these systems $\lambda = \sigma = \sigma_{11} + 2\sigma_{12} + \sigma_{22}$.

III. BINARY SYSTEMS

Of course the conditions for validity of the Nernst-Planck equations are impossible to achieve in general. Tables I and II show some values of D_{ij}, D_i λ, t_i, and σ_{ij} for some particularly simple aqueous binary electrolyte solutions. The data are all from Miller,[13] who reports ℓ_{ij} values for these and many more systems and concentrations. The ℓ_{ij} are strong functions of concentration. Explicit equations for the binary coefficients are

$$D_{11} = \frac{RT}{c_1}\left[\ell_{11}\left(1 + \frac{\partial \ell n y_1}{\partial \ell n c_1}\right) + \ell_{12}\frac{\partial \ell n y_2}{\partial \ell n c_1}\right],$$

$$D_{12} = \frac{RT}{c_2}\left[\ell_{11}\frac{\partial \ell n y_1}{\partial \ell n c_2} + \ell_{12}\left(1 + \frac{\partial \ell n y_2}{\partial \ell n c_2}\right)\right],$$

$$D_{21} = \frac{RT}{c_1}\left[\ell_{21}\left(1 + \frac{\partial \ell n y_1}{\partial \ell n c_1}\right) + \ell_{22}\frac{\partial \ell n y_2}{\partial \ell n c_1}\right], \tag{3.1}$$

$$D_{22} = \frac{RT}{c_2}\left[\ell_{21}\frac{\partial \ell n y_1}{\partial \ell n c_2} + \ell_{22}\left(1 + \frac{\partial \ell n y_2}{\partial \ell n c_2}\right)\right].$$

For binary aqueous uni-univalent electrolytes the Nernst-Planck equations appear to be good approximations up to concentrations of ~ 0.1 molar, but cross terms and non-ideality become increasingly important for higher concentrations. For more complicated salts, the cross effect and non-ideality effects become important at much lower concentrations.

The specific conductance is proportional to concentration, while transference numbers are independent of concentration. Ignoring cross-effects in the specific conductance for these systems is almost 10% incorrect for 0.1 molar solutions. Since solid state ionic conductors are relatively quite concentrated, cross-effects cannot be ignored.

For very dilute solutions, however, it is clear that the Nernst-Planck equations are satisfactory approximations and moreover that all coefficients but λ are essentially constant. Thus, for dilute binary solutions, the transport equations are

$$\frac{\partial E}{\partial t} = \frac{4\pi}{\varepsilon}\left[I + Fz_1D_1\frac{\partial c_1}{\partial x} + Fz_2D_2\frac{\partial c_2}{\partial x} - \frac{F^2}{RT}\left(z_1^2c_1D_1 + z_2^2c_2D_2\right)E\right],$$

$$\frac{\partial c_1}{\partial t} = \frac{\partial}{\partial x}\left(D_1\frac{\partial c_1}{\partial x} - c_1z_1D_1\frac{F}{RT}E\right), \tag{3.2}$$

$$\frac{\partial c_2}{\partial t} = \frac{\partial}{\partial x}\left(D_2\frac{\partial c_2}{\partial x} - c_2z_2D_2\frac{F}{RT}E\right).$$

Contrast these with the full equations, correct for any concentration,

$$\frac{\partial E}{\partial t} = \frac{4\pi}{\varepsilon} [I + F(z_1D_{11} + z_2D_{21}) \frac{\partial c_1}{\partial x} + F(z_1D_{12} + z_2D_{22}) \frac{\partial c_2}{\partial x} - \lambda E],$$

$$\frac{\partial c_1}{\partial t} = \frac{\partial}{\partial x} (D_{11} \frac{\partial c_1}{\partial x} + D_{12} \frac{\partial c_2}{\partial x} - \frac{\lambda t_1}{Fz_1} E) \ , \tag{3.3}$$

$$\frac{\partial c_2}{\partial t} = \frac{\partial}{\partial x} (D_{21} \frac{\partial c_1}{\partial x} + D_{22} \frac{\partial c_2}{\partial x} - \frac{\lambda t_2}{Fz_2} E) \ .$$

The correct equations, Eqs. (3.3), are not appreciably more difficult to solve than the approximate ones, Eqs. (3.2).

Ions 1 and 2 (cation and anion, respectively), distribute themselves quite similarly in space and time because systems are electrically neutral in most regions at most times. Near charged surfaces, such as membranes and electrodes and walls, however, cations and anions separate. To distinguish electrically neutral behavior from non-electrically-neutral behavior, it is useful to introduce the alternative sum (S) and difference (Δ) composition variables defined by[15]

$$S = \frac{1}{2} (\frac{c_2}{z_1} - \frac{c_1}{z_2}) \ , \quad \Delta = \frac{1}{2} (\frac{c_2}{z_1} + \frac{c_1}{z_2}) \ . \tag{3.4}$$

Then

$$c_1 = z_2(\Delta - S) \ , \quad c_2 = z_1(\Delta + S) \ , \tag{3.5}$$

and the transport equations become

$$\frac{\partial E}{\partial t} = \frac{4\pi}{\varepsilon} (I + 2z_1z_2FL_{\Delta\Delta} \frac{\partial \Delta}{\partial x} + 2z_1z_2FL_{\Delta S} \frac{\partial S}{\partial x} - \lambda E) \ ,$$

$$\frac{\partial \Delta}{\partial t} = \frac{\partial}{\partial x} (L_{\Delta\Delta} \frac{\partial \Delta}{\partial x} + L_{\Delta S} \frac{\partial S}{\partial x} - \frac{\lambda}{2z_1z_2F} E) \ , \tag{3.6}$$

$$\frac{\partial S}{\partial t} = \frac{\partial}{\partial x} [L_{S\Delta} \frac{\partial \Delta}{x} + L_{SS} \frac{\partial S}{\partial x} - \frac{\lambda}{2z_1z_2F} (t_2 - t_1) E] \ ,$$

with

$$L_{\Delta\Delta} = \frac{1}{2}(D_{11} + D_{22}) + \frac{1}{2}(\frac{z_1}{z_2} D_{12} + \frac{z_2}{z_1} D_{21}) ,$$
$$L_{\Delta S} = -\frac{1}{2}(D_{11} - D_{22}) + \frac{1}{2}(\frac{z_1}{z_2} D_{12} - \frac{z_2}{z_1} D_{21}) ,$$
$$L_{S\Delta} = -\frac{1}{2}(D_{11} - D_{22}) - \frac{1}{2}(\frac{z_1}{z_2} D_{12} - \frac{z_2}{z_1} D_{21}) , \quad (3.7)$$
$$L_{SS} = \frac{1}{2}(D_{11} + D_{22}) - \frac{1}{2}(\frac{z_1}{z_2} D_{12} + \frac{z_2}{z_1} D_{21}) .$$

The similarity of the coefficients in the first two of Eqs. (3.6) is no accident since the Poisson equation, Eq. (1.4) is, in the new variable Δ,

$$\frac{\partial E}{\partial x} = \frac{4\pi F}{\varepsilon} 2z_1 z_2 \Delta . \quad (3.8)$$

Note that since $(\partial^2 E/\partial t \partial x) = (\partial^2 E/\partial x \partial t)$, Eqs. (3.8) and the first of Eqs. (3.6) require

$$\frac{\partial I}{\partial x} = 0 , \quad (3.9)$$

at least for constant E. This has also been mentioned by Macdonald.[16]

For very dilute solutions, the transformation of Eqs. (3.4)-(3.5) yields, instead of Eqs. (3.2),

$$\frac{\partial E}{\partial t} = \frac{4\pi}{\varepsilon}[I + 2z_1 z_2 F N_+ \frac{\partial \Delta}{\partial x} + 2z_1 z_2 F N_- \frac{\partial S}{\partial x}$$
$$+ 2z_1 z_2 \frac{F^2}{RT}(P_-\Delta + P_+ S)E] , \quad (3.10)$$
$$\frac{\partial \Delta}{\partial t} = \frac{\partial}{\partial x}[N_+ \frac{\partial \Delta}{\partial x} + N_- \frac{\partial S}{\partial x} + \frac{F}{RT}(P_-\Delta + P_+ S)E] ,$$
$$\frac{\partial S}{\partial t} = \frac{\partial}{\partial x}[N_- \frac{\partial \Delta}{\partial x} + N_+ \frac{\partial S}{\partial x} + \frac{F}{RT}(P_+\Delta + P_- S)E] ,$$

with

$$N_+ = \frac{1}{2}(D_1 + D_2) , \quad N_- = -\frac{1}{2}(D_1 - D_2)$$
$$\quad (3.11)$$
$$P_+ = \frac{1}{2}(z_1 D_1 - z_2 D_2) , \quad P_- = -\frac{1}{2}(z_1 D_1 + z_2 D_2) .$$

Again, Eqs. (3.10) are not appreciably less difficult to solve than Eqs. (3.6).

IV. SOLVING THE EQUATIONS

Analytic solutions of Eqs. (3.3), or (3.6), or (3.2), or (3.10) are unlikely in general because the coefficient of E in the right-hand term is explicitly or implicitly proportional to concentration. Moreover, the diffusion coefficients are in general composition dependent. It is nevertheless worthwhile to examine the equation for the simplest case that all coefficients are constants. Poisson's equation itself is then used to obtain, instead of the second two of Eqs. (3.3),

$$\begin{aligned} \frac{\partial c_1}{\partial t} &= D_{11} \frac{\partial^2 c_1}{\partial x^2} + D_{12} \frac{\partial^2 c_2}{\partial x^2} - \frac{t_1}{z_1} \tau_D (z_1 c_1 + z_2 c_2) , \\ \frac{\partial c_2}{\partial t} &= D_{21} \frac{\partial^2 c_1}{\partial x^2} + D_{22} \frac{\partial^2 c_2}{\partial x^2} - \frac{t_2}{z_2} \tau_D (z_1 c_1 + z_2 c_2) . \end{aligned} \tag{4.1}$$

where τ_D, the "dielectric relaxation time", is defined by[15]

$$\tau_D = 4\pi\lambda/\varepsilon . \tag{4.2}$$

The general solutions to these coupled equations can be obtained with standard matrix techniques.[15]

Similarly, the S and Δ equations, Eqs. (3.6), become, for constant coefficients,

$$\begin{aligned} \frac{\partial \Delta}{\partial t} &= L_{\Delta\Delta} \frac{\partial^2 \Delta}{\partial x^2} + L_{\Delta S} \frac{\partial^2 S}{\partial x^2} - \tau_D \Delta , \\ \frac{\partial S}{\partial t} &= L_{S\Delta} \frac{\partial^2 \Delta}{\partial x^2} + L_{\Delta S} \frac{\partial^2 S}{\partial x^2} - \tau_D (t_2 - t_1) \Delta , \end{aligned} \tag{4.3}$$

and these, too, can be uncoupled by standard techniques.[15] The Nernst-Planck versions of Eqs. (4.1) and (4.2) are similar, and are similarly attacked.

The next step in attempting explicit solutions to the equations of Section III could be a perturbation expansion about the constant coefficient case just described. The zeroth-order solution for the experiment in which there is initially a step-function in concentrations has recently been obtained.[15] Zeroth-order solutions for other initial and boundary conditions, for example systems in contact with reservoirs of constant concentration, are important in membrane applications and, presumably, in fuel cell and other energy-producing devices. Perram's recent solutions[17] of the Nernst-Planck analogues of Eqs. (4.1) agree well with experimental membrane results.

In order to obtain maximum information from the equations, computer solution seems necessary because the equations are non-linear. Brumleve and Buck[11] have dealt fully with the numerical solution of the Nernst-Planck type transport equations, but numerical solutions to the more general equations have rarely been attempted.[15,18] Since the more general equations are of essentially the same form as their Nernst-Planck analogues, there seems to be no general reason not to deal only with the correct equations.

The principal difficulty in dealing with either the general equations or the Nernst-Planck equations is lack of data for diffusion, conductance, transference numbers, etc. This difficulty is great for conduction in solids, where hardly any data exists. Both computer simulation and actual experiments must be undertaken to provide estimates of the required parameters.

For membrane applications and battery-type devices, Brumleve and Buck[11] have devised efficacious boundary conditions which presume constant composition reservoirs and first order rate law forms. For each ionic species they write, for $0 \leqslant x \leqslant d$,

$$\begin{aligned} (c_i v_i)_{0,t} &= k_{fL,i}\, c_{iL}(t) - k_{bL,i}\, c_{i0}(t) , \\ (c_i v_i)_{d,t} &= k_{fR,i}\, c_{iR}(t) + k_{bR,i}\, c_{id}(t) , \end{aligned} \qquad (4.4)$$

where $k_{fL,i}$ is the forward rate constant in the left-hand reservoir, c_{iL} is the constant concentration of ion i in that reservoir, $k_{bL,i}$ is the backward rate constant at the left-hand wall, and the other equation refers similarly to the right-hand reservoir and wall. They have also devised very efficient and straightforward grid spacing techniques for both space and time.

V. NONISOTHERMAL ION TRANSPORT

When there is a temperature gradient, Eq. (2.2) becomes

$$-J_i^H = \sum_j \ell_{ij} \frac{\partial_T \mu_i}{\partial x} + \ell_{iQ} \frac{\partial \ell nT}{\partial x} , \tag{5.1}$$

and there is a similar Onsager equation for the flux of heat Q,

$$-Q = \sum_j \ell_{Qj} \frac{\partial_T \mu_j}{\partial x} + \ell_{QQ} \frac{\partial \ell nT}{\partial x} . \tag{5.2}$$

The chemical potential gradient $(\partial_T \mu_j/\partial x)$ is defined by

$$\frac{\partial_T \mu_j}{\partial x} = \frac{\partial \mu_j}{\partial x} + \tilde{S}_i \frac{\partial T}{\partial x} , \tag{5.3}$$

where $\tilde{S}_i$ is partial molar entropy. In brief, the ℓ_{QQ} term in Eq. (5.2) characterizes thermal conduction, the ℓ_{iQ} term in Eq. (5.1) characterizes thermal diffusion (Soret effect), and the ℓ_{Qj} terms in Eq. (5.2) characterize the diffusion thermal effect (Dufour effect). Nonisothermal ionic conduction in solids has previously been mentioned only infrequently.[5,8]

The transport equations, Eqs. (1.6), become

$$\frac{\partial E}{\partial t} = \frac{4\pi}{\varepsilon} (I + F \sum_i \sum_j z_i D_{ij} \frac{\partial c_j}{\partial x} + F \sum_i z_i \ell_{iQ} \frac{\partial \ell nT}{\partial x} - \lambda E) , \tag{5.4}$$

$$\frac{\partial c_i}{\partial t} = \frac{\partial}{\partial x} (\sum_j D_{ij} \frac{\partial c_j}{\partial x} + \ell_{iQ} \frac{\partial \ell nT}{\partial x} - \frac{\lambda t_i}{F z_i} E) .$$

An additional transport equation for the nonisothermal case comes from the energy conservation equation. An excellent approximation for this equation for the mass-transport problems of interest here is[19]

$$\frac{\partial T}{\partial t} = \frac{\tilde{V}}{\tilde{C}_p} \frac{\partial}{\partial x} \ell_{QQ} \frac{\partial \ell nT}{\partial x} , \tag{5.5}$$

where $\tilde{V}$ is the molar volume of the solution and $\tilde{C}_p$ is the molar constant pressure heat capacity.

The thermal conductivity (ℓ_{QQ}/T) is ordinarily essentially constant, so that Eq. (5.5) yields a linear temperature distribution once a thermal steady state is achieved. Since thermal conduction is roughly 100 times faster than diffusion, the temperature gradient terms in Eqs. (5.4) vanish long before the concentration gradient terms. If the time dependence of E is of interest, however, then the time dependence of T must be included through Eq. (5.5).

ACKNOWLEDGMENT

A considerable portion of the research which led to this paper was done at Odense University in July and August, 1979. I gratefully acknowledge both the financial support provided by the European Economic Community Battery Project and the hospitality of Professor J.W. Perram. I also acknowledge with gratitude the careful reading of this paper by Bruce Borey.

REFERENCES

1. L. Onsager, Phys. Rev. 37, 405 (1931); 38, 2265 (1932).
2. J. Meixner and H.G. Reik, in: "Handbuch der Physik," Vol. III/2, S. Flügge, ed., Springer-Verlag, Berlin (1959); D.G. Miller, Chem. Rev. 60, 15 (1960); S.R. de Groot and P. Mazur, "Non-Equilibrium Thermodynamics," North-Holland, Amsterdam (1962); D.D. Fitts, "Nonequilibrium Thermodynamics," McGraw-Hill, New York (1962); R. Haase, "Thermodynamics of Irreversible Processes," Addison-Wesley, Reading, MA (1963); A. Katchalsky and P.F. Curran, "Nonequilibrium Thermodynamics in Biophysics," Harvard, Cambridge, MA (1965); D.G. Miller, in "Proc. Int. Symp. on Foundations of Continuum Thermodynamics," J.J. Domingos, M.N.R. Nina, and J.H. Whitelaw, eds., Wiley, New York (1975).
3. Opponents of Onsager thermodynamics are led by C. Truesdell. See, in particular, his "Rational Thermodynamics," McGraw Hill, New York (1969).
4. R.P. Buck, in "Critical Reviews in Analytical Chemistry," Chemical Rubber Company, Cleveland (1979).

5. R.E. Howard and A.B. Lidiard, Rep. Prog. Phys. 27, 161 (1964).
6. T.R. Anthony, in "Diffusion in Solids: Recent Developments," A.S. Nowick and J.J. Burton, eds., Academic Press, New York (1965), 353; W.J. Fredericks, ibid, 381.
7. C. Wagner, Progr. Solid State Chem. 10, 3 (1975).
8. J.C. Kimball, Phys. Rev. B16, 785 (1977).
9. G.J. Dudley and B.C.H. Steele, J. Solid State Chem. 31, 233 (1980).
10. H. Cohen and J.W. Cooley, Biophys. J. 5, 145 (1965).
11. T.R. Brumleve and R.P. Buck, J. Electroanal. Chem. 90, 1 (1978).
12. L. Duckworth, "Electricity and Magnetism," Holt, Rinehart and Winston, New York (1960), p. 96.
13. D.G. Miller, J. Phys. Chem. 70, 2639 (1966).
14. R.A. Huggins and J.M. Huggins, Macromolecules 10, 889 (1977).
15. J.H. Leckey and F.H. Horne, J. Phys. Chem. (to be published) (1981).
16. J.R. Macdonald, J. Appl. Phys. 46, 4602 (1975).
17. J.W. Perram, Comm. Math. Chem. 1980, 37 (1980).
18. R.M. Goldberg and H.S. Frank, J. Phys. Chem. 76, 1758 (1972).
19. R.L. Rowley and F.H. Horne, J. Chem. Phys. 72, 131 (1980).

ION TRANSPORT BOUNDARY CONDITIONS

Frederick H. Horne,† John H. Leckey,† and
John W. Perram‡

†Department of Chemistry
Michigan State University
East Lansing, MI 48824 U.S.A.

‡Matematisk Institut
Odense Universitet
DK 5230, Odense, Denmark

In theoretical studies of ion transport in heterogeneous systems it is convenient to employ kinetic boundary conditions of the form[1,2]

$$j_i = k_{if}c_{i0} - k_{ib}c_{i\delta} , \tag{1}$$

for the flux j_i of ion i through an interface of thickness δ.

Interfacial Region

The forward and reverse rate constants are k_{if} and k_{ib} (the directions are indicated in the drawing), and c_i is the ion concentration. The purpose of this article is to determine the assumptions required for validity of boundary conditions of this type.

At equilibrium, the fluxes vanish and the Nernst Distribution Law[3] obtains for the ratio of activities a_i in the absence of an electric field,

$$\frac{a_{i\delta}}{a_{i0}} = \frac{c_i y_{i\delta}}{c_{i0} y_{i0}} = K_{iN} = \exp\left(\frac{-\Delta G_i^{\ominus}}{RT}\right) , \tag{2}$$

where y_i is the molarity based activity coefficient, K_{iN} is the Nernst Distribution Constant, and the standard Gibbs free energy difference is the difference in standard chemical potential across the interface,

$$\Delta G_i^{\ominus} = \mu_{i\delta}^{\ominus} - \mu_{i0}^{\ominus} . \tag{3}$$

If an electric potential difference, $\Delta\phi = \phi_\delta - \phi_0$, is maintained across the interface, then Eq. (2) becomes

$$(a_{i\delta}/a_{i0}) = K_{iN} \exp(-Z_i F\Delta\phi/RT) . \tag{4}$$

Equations (2)-(4) are readily derived from the equilibrium requirement that $\mu_{i\delta} = \mu_{i0}$, with the electrochemical potential given by

$$\mu_i = \mu_i^{\ominus} + RT\ell n a_i + z_i F\phi . \tag{5}$$

If we put $j_i = 0$ in Eq. (1), we find

$$(c_{i\delta}/c_{i0}) = (k_{if}/k_{ib}) . \tag{6}$$

Thus, by Eq. (2), $(k_{if}/k_{ib}) = K_{iN}$ for ideal solutions at equilibrium in the absence of an electric field and, by Eq. (4), $(k_{if}/k_{ib}) = K_{iN} \exp(-z_i F\Delta\phi/RT)$ for ideal solutions at equilibrium in the presence of an electric field.

If we neglect cross terms, or if only one ion is flowing, the flux of an ion in one dimension is given by[4]

$$-j_i = \ell_{ii} \frac{d\mu_i}{dx} , \tag{7}$$

where ℓ_{ii} is the Onsager coefficient and μ_i is the electrochemical potential. The electrochemical potential gradient is, by Eq. (5),

$$\frac{d\mu_i}{dx} = \frac{RT\Gamma_i}{c_i}\frac{dc_i}{dx} + z_i F\frac{d\phi}{dx} + \frac{d\mu_i^{\ominus}}{dx}, \tag{8}$$

with

$$\Gamma_i = 1 + \frac{\partial \ln y_i}{\partial \ln c_i}. \tag{9}$$

Substitution of Eq. (8) into Eq. (7) yields

$$-j_i = D_i\frac{dc_l}{dx} + \frac{z_i F D_i c_i}{RT}\frac{d\phi}{dx} + \frac{D_i c_i}{RT}\frac{d\mu_i^{\ominus}}{dx}, \tag{10}$$

where[4]

$$D_i = \frac{RT\ell_{ii}}{c_i}, \tag{11}$$

and where we have neglected the activity coefficient term, which is consistent with the neglect of cross-terms.[4]

Eq. (10), rearranged, is a differential equation for c_i,

$$\frac{dc_i}{dx} + \frac{c_i}{RT}\frac{dA_i}{dx} = \frac{-j_i}{D_i}, \tag{12}$$

with

$$A_i \equiv \mu_i^{\ominus} + z_i F\phi. \tag{13}$$

If A_i is linear in x and j_i and D_i are constants across the interface, then the solution of Eq. (12) is

$$c_i = c_{i0}\exp\left(-\frac{x\Delta A_i}{\delta RT}\right) + j_i\frac{\delta RT}{D_i\Delta A_i}\left[\exp\left(-\frac{x\Delta A_i}{\delta RT}\right) - 1\right], \tag{14}$$

with

$$\Delta A_i = \Delta G_i^{\ominus} + z_i F\Delta\phi \ . \tag{15}$$

The interfacial boundary condition, Eq. (1), is obtained by putting $x = \delta$ in Eq. (14) and solving for j_i. This results in

$$k_{if} = \frac{D_i \Delta A_i}{\delta RT} [\exp(-\frac{\Delta A_i}{RT})][- \exp(-\frac{\Delta A_i}{RT})]^{-1} \ ,$$

$$k_{ib} = \frac{D_i \Delta A_i}{\delta RT} [1 - \exp(-\frac{\Delta A_i}{RT})]^{-1} \ . \tag{16}$$

The ratio of k_{if} to k_{ib} is

$$\frac{k_{if}}{k_{ib}} = \exp(-\frac{\Delta\mu_i^{\ominus} + z_i F\Delta\phi}{RT}) \ , \tag{17}$$

as required by Eqs. (4) and (6) for ideal solutions.

If z_i and $\Delta\phi$ are positive and if $z_i F\Delta\phi > \Delta\mu_i^{\ominus}$, then $k_{if} < k_{ib}$. This is as expected since a positive ion would be moving into a region of higher electric potential as it moves in the positive x direction. Similarly, if z_i is negative, $\Delta\phi$ is positive, and $z_i F\Delta\phi > \Delta\mu_i^{\ominus}$, then $k_{if} > k_{ib}$. This is also as expected. The other term in the exponent of Eq. (17), $\Delta\mu_i^{\ominus}$, can be approximated by the not very accurate Born equation,[2]

$$\Delta\mu_i^{\ominus} = \frac{z_i^2 F^2}{2a_i} (\frac{1}{\varepsilon_\delta} - \frac{1}{\varepsilon_0}) \ , \tag{18}$$

where a_i is the radius of the ion and ε_1 and ε_2 are the dielectric coefficients of the 2 media. The rate constant corresponding to transfer to the higher dielectric medium will be higher for $\Delta\phi = 0$. Clearly the parameter δ, the thickness of the interface, is not readily accessible. It may be possible to estimate it from experiments which determine k_{if}, k_{ib} and the distribution coefficient K_{iN} for the ion between the two phases.

In conclusion, Eq. (11) is valid if: (i) the Nernst-Planck equations are obeyed in the interface,[4] (ii) $\mu_i^{\ominus}$ and ϕ are linear functions of x, and (iii) j_i and D_i are constants across the interface.

Improvements in Eq. (1) due to cross-effects and non-ideality will be the goal of further research. The forms of Eqs. (16) are not likely to be significantly affected by any nonlinear behavior of $\mu_i^{\ominus}$ and ϕ across the interface. This, too, should receive further attention.

REFERENCES

1. F.R. Brumleve and R.P. Buck, J. Electroanal. Chem., 90, 1 (1978).
2. J.W. Perram, J. Math. Chem., 8, 37 (1980).
3. I. Prigogine and R. Defay, "Chemical Thermodynamics", D.E. Everett, trans., Longmans, London (1954).
4. F.H. Horne, this volume. (Ch.14).

INDEX